Microsoft Excel® Manual

to accompany

Elementary Statistics

A Step by Step Approach

Seventh Edition

Allan G. Bluman
Professor Emeritus
Community College of Allegheny County

Prepared by
Renee Goffinet
Spokane Community College

Virginia Koehler

Ron Merchant
University of Phoenix

Updated by Aziel Wilson
North Central Texas College

Boston Burr Ridge, IL Dubuque, IA New York San Francisco St. Louis
Bangkok Bogotá Caracas Kuala Lumpur Lisbon London Madrid Mexico City
Milan Montreal New Delhi Santiago Seoul Singapore Sydney Taipei Toronto

The McGraw-Hill Companies

Microsoft Excel® Manual to accompany
ELEMENTARY STATISTICS: A STEP BY STEP APPROACH, SEVENTH EDITION
BLUMAN, ALLAN

Published by McGraw-Hill Higher Education, an imprint of The McGraw-Hill Companies, Inc., 1221 Avenue of the Americas, New York, NY 10020.

This book is printed on recycled, acid-free paper containing 10% post consumer waste.

1 2 3 4 5 6 7 8 9 0 QSR/QSR 0 9 8

ISBN: 978-0-07-333122-5
MHID: 0-07-333122-8

www.mhhe.com

TABLE OF CONTENTS

CHAPTER

CHAPTER 1

INTRODUCTION TO MICROSOFT EXCEL®

Welcome to the **Excel® Manual for use with Elementary Statistics A Step by Step Approach Seventh Edition** for Excel® 2007.

This manual is a step-by-step approach to using Excel®, the most popular spreadsheet program in the world, to handle a wide variety of statistical applications. This manual will assist you in using Excel® as a powerful tool to analyze numerical data, solve problems, and deliver a visually meaningful product.

Spreadsheets are used to help analyze numerical data and solve problems. Using a spreadsheet, when you enter data you immediately see the results of the related changes; you can run statistical analysis and create charts and graphs to assist in decision making. Using "what if" questions and the Excel® program you can analyze data in a way that would be very time consuming if not impossible without the use of a similar tool. However, you as a student are still responsible for understanding and interpreting each statistical concept. The complete results needed for decision making do not appear magically on the computer.

Chapter 1 is an introduction into the basic functions of Excel®. We do not assume any previous experience working with spreadsheets. After completing this chapter, you will be able to:

- Understand why Excel® is so useful as a statistical tool.
- Create a workbook, or modify an existing workbook.
- Create formulas and solve problems.
- Use a spreadsheet to experiment with and illustrate statistical concepts.

To begin, be sure that the software is loaded on the system you are using. Although these instructions are intended to be used with the Excel® 2007 version of Excel®, they can also be used with earlier versions. While the screens and commands will vary, the general information will remain much the same. If you are using an earlier version, just be flexible with your expectations of the screen view and menu placement. It is also important to note to novice computer users that even if you are using the latest software, every computer system is set up a little differently; hence, there will be minor differences in how things look among Excel® 2007 users.

Launching Microsoft Excel®. If you are not familiar with the system on which you are working, you may need to ask how the system has been set up. You may see an Excel® icon located on your desktop, taskbar or shortcut bar. Using the mouse, point at this icon and left click your mouse button once to open Excel®. If an icon is not located in any of these places, you can try selecting the Start button at the bottom left hand corner of the screen, and selecting Programs. Some systems have a Quick Preview loaded that introduces Microsoft Excel®. It is highly recommended that you go through this if you are not familiar with workbooks and Microsoft Excel®.

As you work through this manual, you will find that specific items to be selected will be differentiated with the use of an alternate font style. For example, note in the previous paragraph that the words Excel® icon, Start button, and Programs appear in a different font. Text that you are required to key into the spreadsheet will be noted by the bold style in addition to the differing font. Once Microsoft Excel® has loaded, a file called a *workbook* automatically opens.

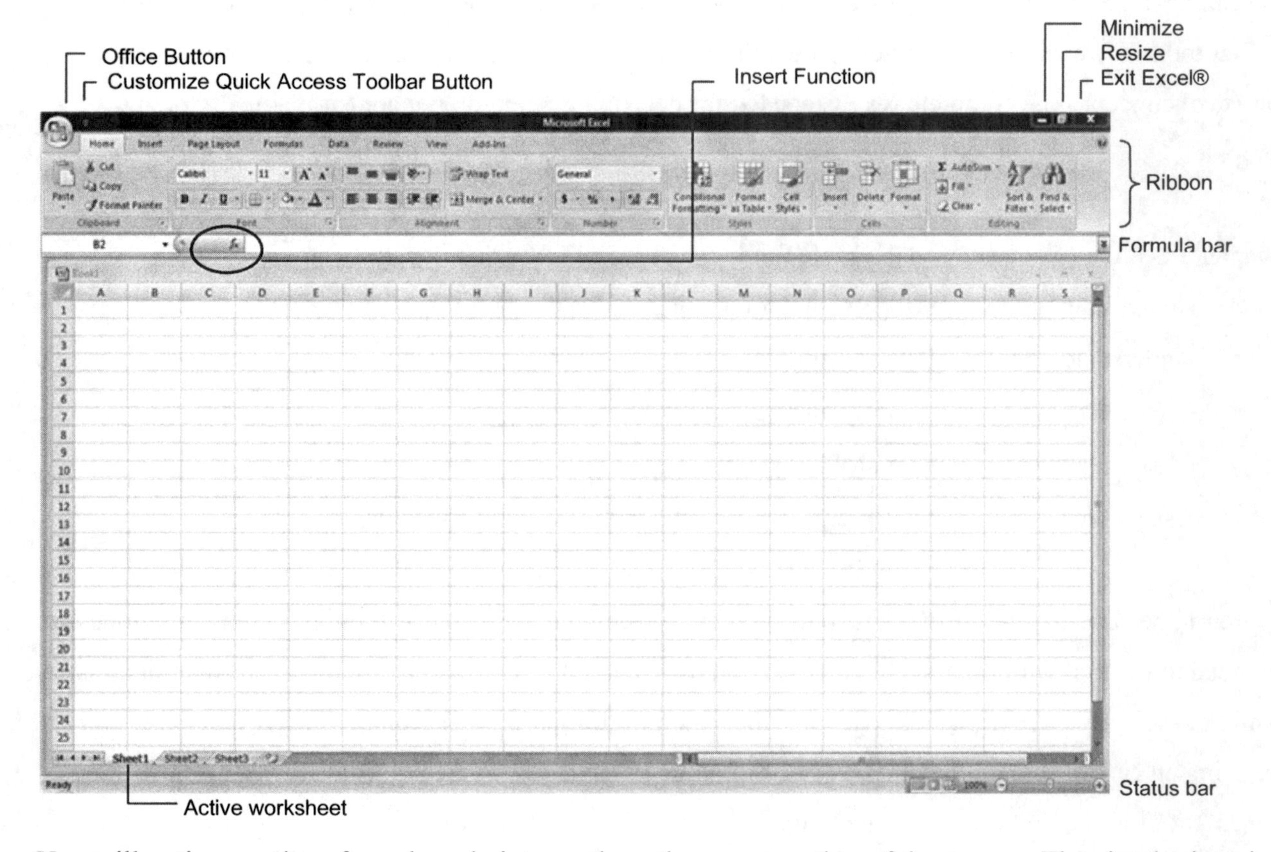

You will notice a series of words and pictures along the upper portion of the screen. The circular icon is called the Office Button. The drop down arrow to the right is the Quick Access Toolbar. The collection of words and icons below these two items is called the Ribbon. The Ribbon will change to display different menus according to the tab you select at the top. The formula bar contains the Insert Function icon.

The Microsoft Excel® 2007 Ribbon

If you have used past versions of Excel®, the first thing that you will notice is instead of the menu bar and commands there is a *Ribbon*. The Ribbon contains *tabs* designed to organize commands by related activities. On each tab you will notice *groups*. The groups consist of collections of related command icons for editing the workbook.

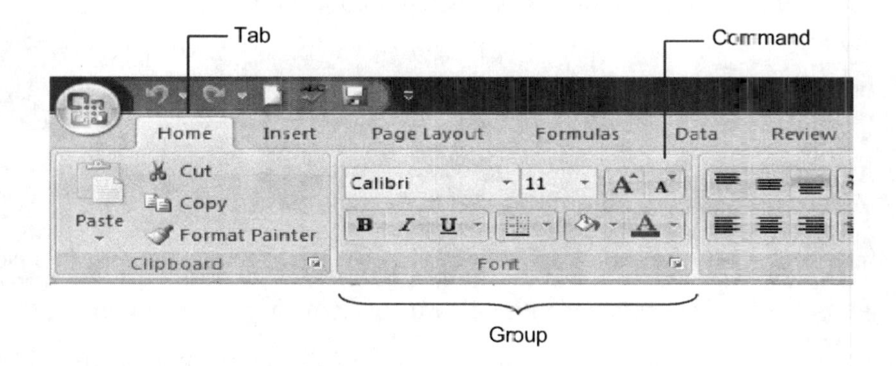

Navigating Excel®

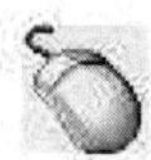

You can use your mouse to select an object, function, or launch an action. As you move your pointer around the screen, it will alternate between an arrow, and an open plus sign .

When your mouse pointer rests on an icon on the Ribbon, a short description is displayed just below that icon.

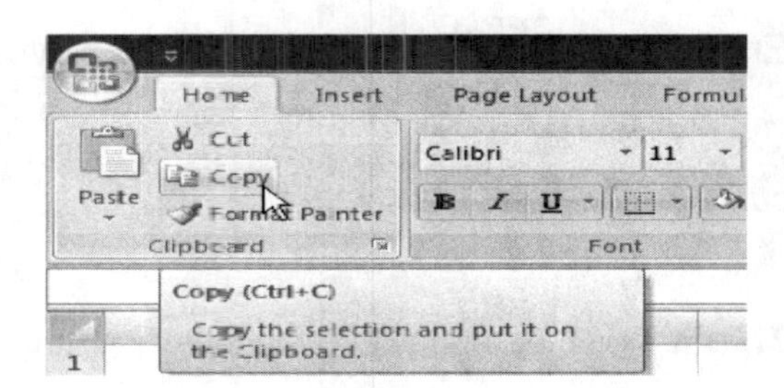

You can make commands quickly available by adding them to the Quick Access toolbar.

1. Click on the Customize Quick Access Toolbar Button and select the New command.
2. Repeat step 1 to select the Undo, Redo, and Spelling and Grammar commands.

Create a New Workbook

If a blank workbook does not appear, you will need to create a new workbook.

To create a new workbook, select the Office Button, from the Menu bar. Select New. If you have added the New icon to the Quick Access Toolbar, you may choose to click it instead.

A workbook contains one or more *worksheets*. Each worksheet consists of rows, columns and cells. Each individual rectangle is a *cell*. Each cell is identified by its placement in the Column (A, B, C ...) and the Row (1, 2, 3 ...). Thus, the cell B3 would be in the 2nd column and the 3rd row. The mouse pointer in Excel® looks like an open plus sign. When the pointer is in a cell, click the left mouse button and that cell becomes the active cell. The cell will have a dark bordered box around it. You may also use the arrow keys: up, down, right, left, to move around in the worksheet.

The data or information you key will show in the active cell and in the formula bar. When you press the Enter key, your data is entered into the cell and the cell immediately below becomes the active cell. Or you can point to another cell or use your arrow keys. Pressing the tab key will move the curser to the right, activating that cell.

To edit a cell, double click your mouse pointer in that cell, and the cell can be edited. The mouse pointer will show as a large I beam instead of an open plus.

You may then edit the cell, without re-keying the entire contents.

You can select several cells at once to work with, called a range. A *range* is a rectangular group of cells. To use your mouse in selecting a range place your mouse pointer on the upper left cell of the range to be highlighted, then click, hold and drag your mouse pointer to the lower right cell of the range, and release the mouse button. The first cell shows a white background, all other cells in the selected range show a gray background. A range is identified by its first and last cells with a colon in between.

Suppose you wanted to show the distribution of the number of pounds of snack food consumed during a sporting event.

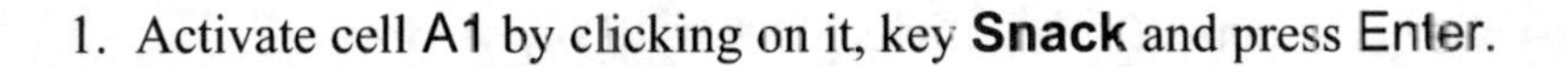

1. Activate cell A1 by clicking on it, key **Snack** and press Enter.

2. Cell A2 should now be active. Key **Potato chips** and press Enter.

3. In cell A3 key **Tortolla chips**. (You will correct the spelling later.) Press Enter.

Enter the remaining data as follows.

Notice that the words do not all fit into one cell and go into the adjacent cell. When you type a number into **B2**, you will no longer be able to see all of the text in **A2**.

Snack	Pounds
Potato Chips	11.2
Tortolla Chips	8.2
Pretzels	4.3
Popcorn	3.8
Snack Nuts	2.5

If you make an error, you can correct it by immediately by selecting the Undo icon from the Quick Access Toolbar bar.

You will now widen column A. To widen and automatically fit the column width, do the following.

1. Place the mouse pointer in the column headings row between columns A and B. The mouse pointer changes to a thick, black plus sign

example chp1

	A	B
1	Snack	Pounds
2	Potato Ch	11.2
3	Tortolla Cl	8.2
4	Pretzels	4.3
5	Popcorn	3.8
6	Snack Nut	2.5

2. Double click the left mouse button. The column automatically widens.

To move text to a new position the Home tab must be selected

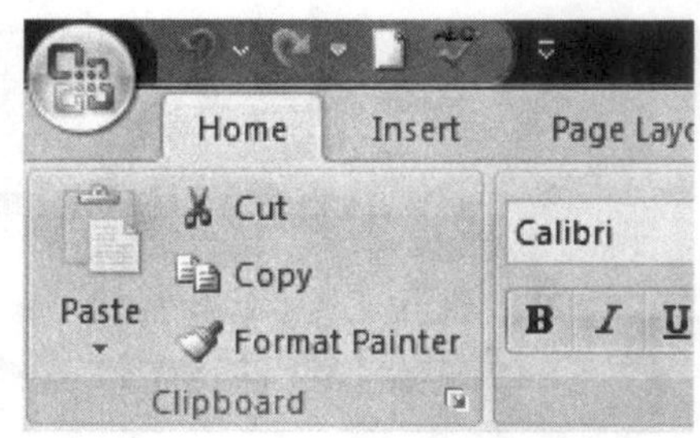

1. Highlight the cells in the range **B1:B6**, click on the Cut icon.

2. Select the new location by activating the first cell in which you wish the text to be located **C1**, click on the Paste icon.

To copy text, follow the same procedures for moving text, but instead of the Cut icon, select the Copy icon.

Now you will practice the cut and paste procedure by moving the numbers in Column C to Column B.

1. Select the range of cells, **C1:C6**. To do this, click your mouse pointer on cell **C1**. Then Click, hold and drag your mouse pointer to cell **C6**.

2. Click on the Cut icon. The cells have a running black line around them.

3. Click on cell **B1**.

4. Click on the Paste icon. Press the Enter key.

Check Spelling and Grammar

Excel® allows you to check the spelling of the text in your entire workbook or just selected cells.

1. To check the spelling of the whole workbook click on cell **A1**, then click on the Spelling and Grammar icon located under the Review tab or in the Quick Access Toolbar.

2. A dialogue box appears giving you several choices. If the spelling is incorrect, highlight the correct spelling from the suggestions box and click on the Change button.

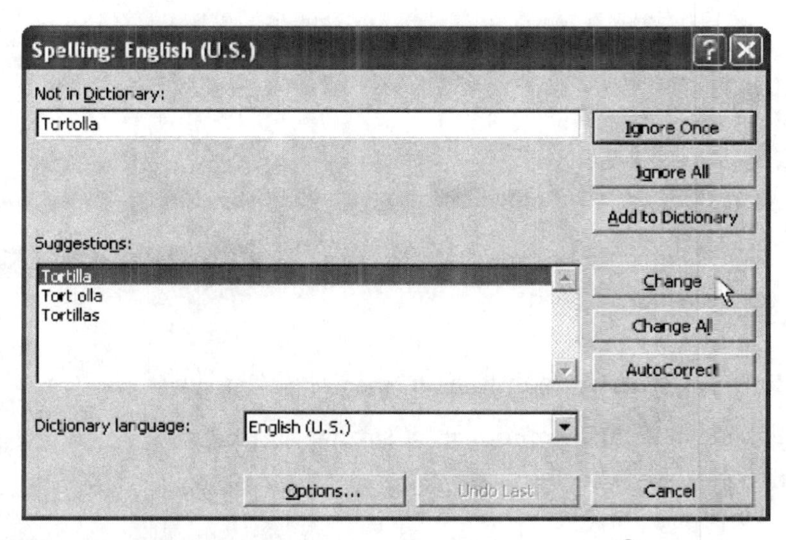

3. To check a selected area, highlight the cell or cells you want Excel® to check, click on the Spelling and Grammar icon, and follow the same procedure as above.

Note: If the spellchecker lands on a word that it does not recognize but is correct, for example a name or abbreviation, you can choose to ignore the suggested change by selecting Ignore Once, or Ignore All or you may add the word to your dictionary by selecting the Add to dictionary button.

Setting Page Orientation

You may create your workbook in either portrait or landscape orientation by doing the following:

1. Select the Page Layout tab on the Ribbon

2. Click on the Orientation and select Portrait or Landscape.

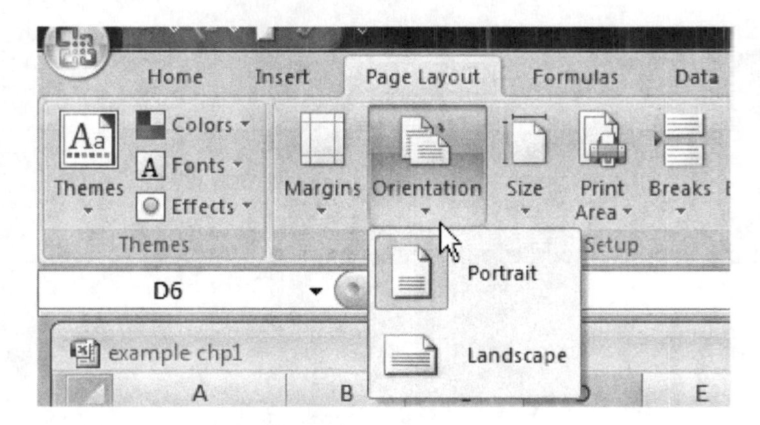

NOTE: As you work through this text, some illustrations may appear different than your screen.

Saving a Workbook

1. Place your USB drive into the USB port (or whatever storage device you are using).

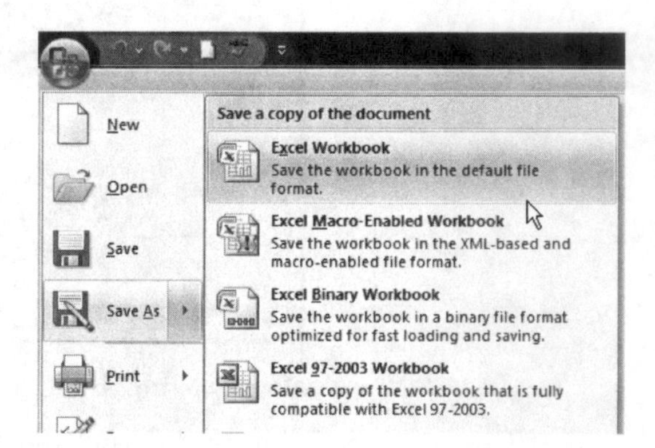

2. Click on the Office Button.

3. Highlight the Save As icon from the menu and click on your choice of format for saving the file

You will see various options for saving your file. For example, the Excel® Workbook option saves the file in the Excel® 2007 format and the Excel® 97-2003 Workbook option saves it to be compatible with earlier Excel® versions.

4. Select the location in which to save your file. Click on the storage device you are using.

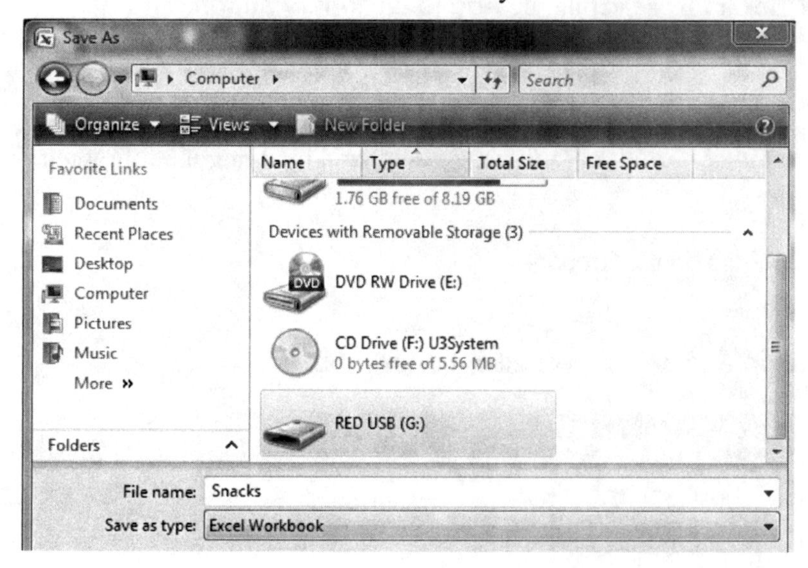

5. Key in the desired file name in the File name text box, use **Snacks**.

6. Click on Save.

Closing a Workbook

To close a workbook (file) do the following:

1. Click on the Office Button.

2. Click the Close icon from the menu.

If no changes have been made since the last save, the file will close. If the current information has not been saved you may select Yes to save the changes, select No if you do not want to save the changes, or select Cancel to return to the current workbook.

Opening a Workbook

Once you have saved and closed a workbook you may need to retrieve it to continue working.

1. Click the Office Button. Click the Open icon from the menu. If you have added it to the Quick Access Toolbar you may click the Open icon on the toolbar instead.

3. If the USB device (or other storage device) does not show, browse to find the device or location where the file was saved. Click to open the device or folder.

4. Click on the file name of the workbook you want to open. (Select Snacks if you closed the workbook earlier).

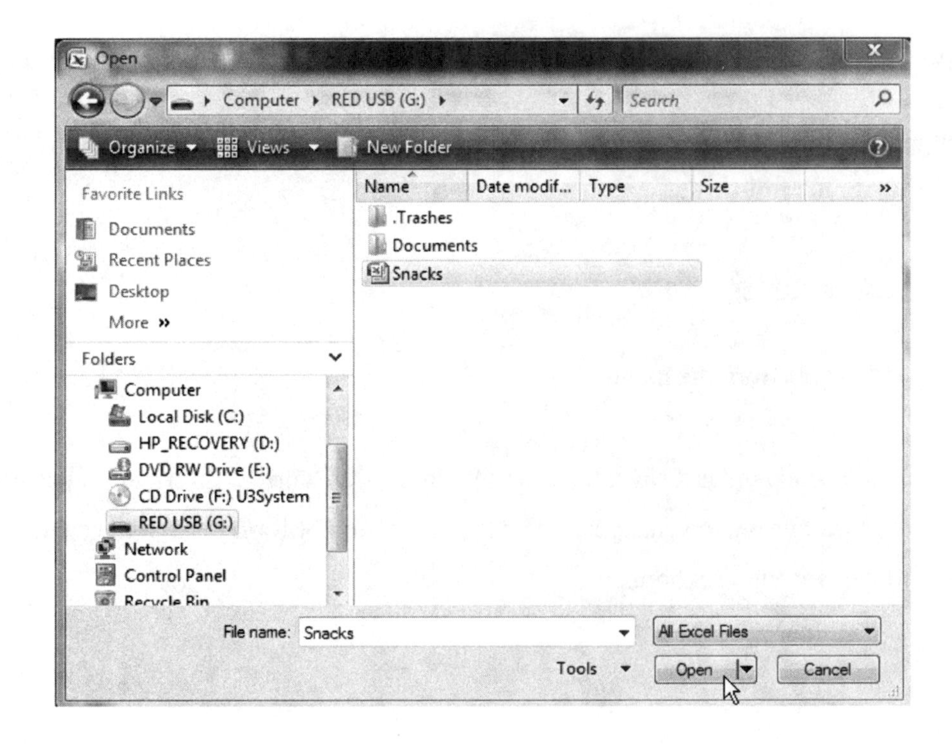

5. Click on Open.

Using Excel® to Solve problems

You will use an existing workbook to create formulas and solve problems.

The following problem is an example of how one can use the workbook. All of the icons for this exercise are located under the Home tab.

Open the file, **Snacks**, if it is not already open. We are interested in finding what percent each snack type is of the total. First you will add the sum of all the values.

1. Make B7 your active cell. Key in =**sum(B2:B6)**. Press the Enter key.

Important Note: To distinguish a formula from text as you enter information on the Excel® worksheet, start **all** formulas with an equal " =" sign.

2. In cell C1, key **% of total.**

3. In cell C2, key =**B2/30** press Enter.

4. Make cell C2 the active cell. Put your mouse pointer on the lower right corner of the cell. It will show a small black box called a handle. The mouse pointer on the handle will show as a thick, black plus. Click your mouse button, hold and drag the mouse pointer down to cell C6.

	A	B	C
1	Snack	Pounds	% of total
2	Potato chips	11.2	0.373333
3	Tortilla chips	8.2	0.273333
4	Pretzels	4.3	0.143333
5	Popcorn	3.8	0.126667
6	Snack Nuts	2.5	0.083333
7		30	

Cells C3 to C6 have automatically been filled in with the formula.

5. Cells C2:C6 should already be highlighted. If not highlight C2:C6. From the Home Tab, click on the Sum function icon. Excel® automatically enters the sum of the highlighted cells immediately below them.

6. With Cells C2:C7 still highlighted, click on the Percent Style icon to show the amounts in percent form instead of decimal form.

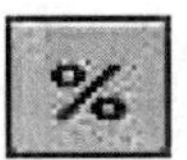

7. Keeping the cells highlighted, click once on the Increase Decimal icon to show the percents to one decimal place.

	A	B	C
1	Snack	Pounds	% of total
2	Potato chips	11.2	37.3%
3	Tortilla chips	8.2	27.3%
4	Pretzels	4.3	14.3%
5	Popcorn	3.8	12.7%
6	Snack Nuts	2.5	8.3%
7		30	100.0%

Rearranging columns

Sometimes you want to rearrange the columns. If you want to show a graph of the percents you would use columns A and C. To move the column C (% of total) next to the column A (Snack), do the following.

1. Right click your mouse pointer anywhere in column B.
2. Select Insert.
3. Choose Entire column.

The information in columns B through C has been shifted to the right creating a blank column.

	A	B	C	D
1	Snack		Pounds	% of total
2	Potato chips		11.2	37.3%
3	Tortilla chips		8.2	27.3%
4	Pretzels		4.3	14.3%
5	Popcorn		3.8	12.7%
6	Snack Nuts		2.5	8.3%
7			30	100.0%

10. Highlight cells D1 to D7.

11. From the Home tab select Cut.

12. Make cell B1 your active cell and select the Paste icon.

The columns with % of total are now next to the column Snack.

You can also select non-adjacent columns, simply highlight one of the columns, and hold down the <Ctrl> key while highlighting another column. This can be done with as many as you wish.
If you wish, save your workbook as **Ch1**. Close your workbook.

Exiting Excel®

1. Click the Office Button.
2. Choose X Exit Excel®

You can also exit Excel® by clicking on the Exit Excel® button located at the top right corner of the Excel® window.

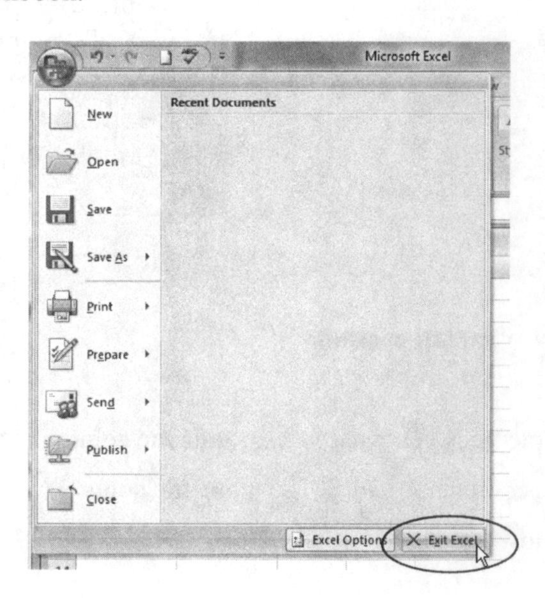

Using Data Sets

Your instructor may assign additional work requiring you to access data sets from either a CD or the publisher's website. If your book was packaged with a CD, the data sets are available on it. Otherwise, the data sets can be found on the book's website. To access the datasets online, follow these instructions:

1. Visit the website www.mathzone.com.
2. Select the Register for Mathzone Now button. If you have used the site before, select Login to your course and enter your username and password. Go to Step 5.
3. Select Bluman: Elementary Statistics 6e from the pull down menu.
4. If you have not used the site before, click First Time User. Then select, I am a Student. Follow the registration instructions.

If you bought a used text and your registration code does not work (or was not included with a used book), the previous book owner may have used the code. You will need to purchase access to the website by clicking "I do not have a registration code" and following the on-screen instructions.

5. After you login, you will need to find the link to the data sets. Click on the Self Study tab.
6. Scroll to the bottom of the list of chapters, select the plus sign beside More Resources and click Data Sets.
7. Click the link Data Sets that will appear.
8. The data sets are contained in zip files. You will see the links for downloading the zip files at the bottom of the page. Click the appropriate format to begin the download.

Please note that the naming convention for the individual data files within the zip file appears above the links.

Notes:

CHAPTER 2
FREQUENCY DISTRIBUTIONS AND GRAPHS

We have all heard that a picture is worth a thousand words. With the use of graphs and charts, a picture is worth a thousand numbers. Using spreadsheet software such as Excel® to generate results of numerical data, inquiry is much easier than most other methods. However, the results are often not easy to visualized or interpret. Even when they are carefully laid out in report form the results are often cumbersome because there is simply too much information for most people to take in. To make it easier to describe situations, draw conclusions, or make inferences about events it helps to organize data into frequency distributions or graphs.

With the use of graphs and charts, the casual user can quickly evaluate the results of the data. The visual presentation allows the viewer to assess the information gathered quickly and usually more accurately than by reviewing the numbers alone. Pie charts show the relationship of parts to a whole. Bar charts show comparison between items or comparison over time. Line charts are often best for showing the amount of change in values over time. As you work through this chapter, take time to experiment with the different types of charts and graphs available to see which best illustrates your point. You can rearrange your data, even after you have charted it or added additional data.

The graphic presentations you see extensively in newspapers including *USA Today*, magazines and governmental reports often portray data from a frequency distribution in the form of pie charts, bar charts, histograms and line charts. With Microsoft Excel®, you can easily turn your data into dynamic graphic presentations. Chapter 2 shows you how to:

- Conveniently organize data into a frequency distribution.
- Utilize the power of Excel® to create such common graphic presentations as frequency distributions, histograms, Pareto charts, time series and pie charts.
- Format the appearance of your chart or graph.
- Enrich a written document by embedding a chart into a report.

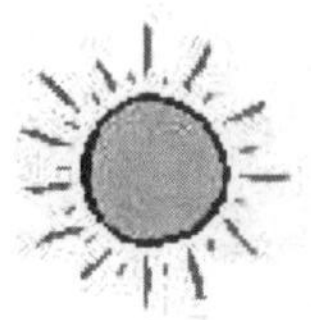

Creating Frequency Distributions using the data files provided

Problem 2-1. The following data represent the record high temperatures for each of the 50 states. Construct a grouped frequency distribution for the data using seven classes.

Excel®'s Histogram Data Analysis feature can be used to construct a frequency distribution. You will key the temperatures of the 50 states in a single column from A2 to A51.

1. In a new worksheet, key **TEMPERATURES** in cell A1.

2. Key the data in a single column into A2 to A51.

112	**100**	**127**	**120**	**134**	**118**	**105**	**110**	**109**	**112**
110	**118**	**117**	**116**	**118**	**122**	**114**	**114**	**105**	**109**
107	**112**	**114**	**115**	**118**	**117**	**118**	**122**	**106**	**110**
116	**108**	**110**	**121**	**113**	**120**	**119**	**111**	**104**	**111**
120	**113**	**120**	**117**	**105**	**110**	**118**	**112**	**114**	**114**

The title is in A1 and the data are displayed in cells A2:A51.

3. In column B, key the Bin numbers (which are the upper limit of each class used in this instance.) In B2 to B8, key the data **104, 109, 114, 119, 124, 129, 134**

Refer to your textbook if necessary for instructions on finding class limits.

4. Save the file on your storage device as **high temps**. You will use this file again later.

5. Select the Data Tab on the Ribbon. Select Data Analysis.

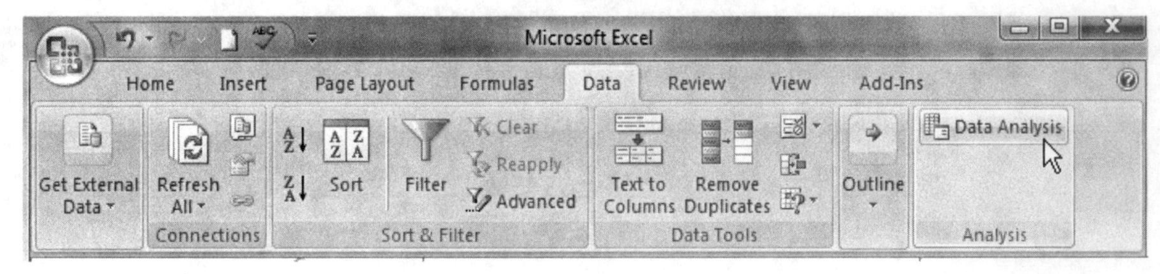

If Data Analysis does not appear, you will need to activate the Data Analysis Toolpak.

1. Click the Office Button.
2. Click the Excel® Options button located at the bottom right of the window.

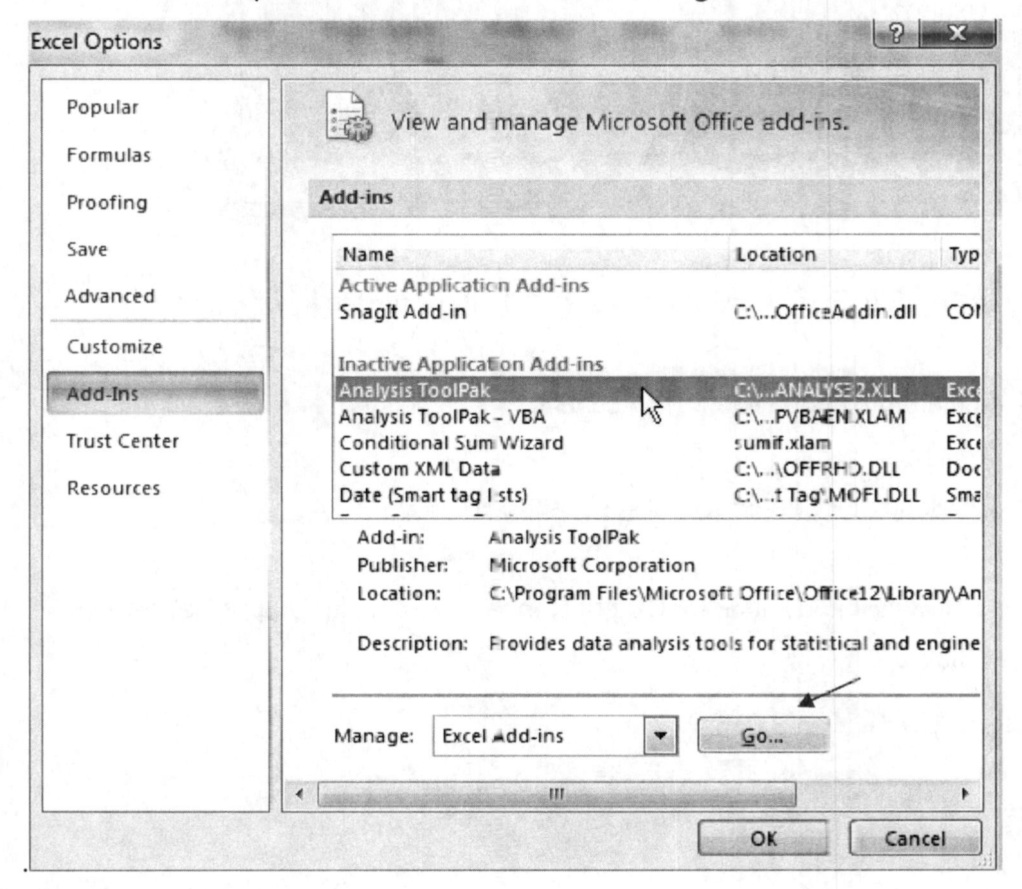

3. Select Add-Ins.
4. Select Analysis Toolpak. Click Go.
5. On the Add-Ins Screen, choose the Analysis Toolpak option. Click OK.
6. Once the Analysis ToolPak option is available, select Data Analysis from the Ribbon. Select Histogram. Click OK.

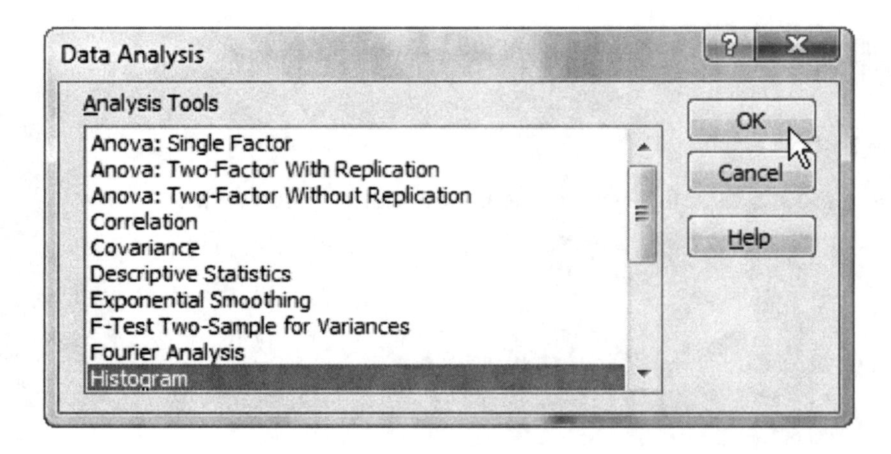

7. In the Input Range textbox, key **A2:A51**. Press the tab key.

8. In the Bin Range textbox, key **B2:B8.**

9. Check box for Labels should not be selected.

10. Select the radio button for Output Range. Click on the Output Range textbox. Key **D1**. Click OK.

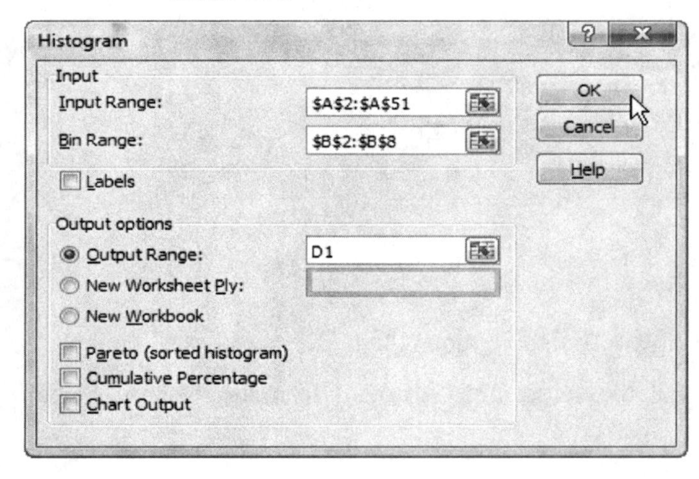

The frequency distribution has been determined for each bin.

	A	B	C	D	E
1	TEMPERATURES			*Bin*	*Frequency*
2	112	104		104	2
3	110	109		109	8
4	107	114		114	18
5	116	119		119	13
6	120	124		124	7
7	100	129		129	1
8	118	134		134	1
9	112			More	0
10	108				

To see the difference when Bin numbers are not assigned, do the following.

11. Select Tools, Data Analysis, and Histogram once more.

12. The information is still displayed. Delete the range of B1:B8 in the Bin Range textbox.

13. In the Output Range textbox, key **G1** Select OK.

	A	B	C	D	E	F	G	H
1	TEMPERATURES			*Bin*	*Frequency*		*Bin*	*Frequency*
2	112	104		104	2		100	1
3	110	109		109	8		104.8571	1
4	107	114		114	18		109.7143	8
5	116	119		119	13		114.5714	18
6	120	124		124	7		119.4286	13
7	100	129		129	1		124.2857	7
8	118	134		134	1		129.1429	1
9	112			More	0		More	1
10	108							

You can see the difference between determining your own class widths and letting Excel® determine them.

14. Name your worksheet by double clicking on Sheet1 the active worksheet tab located at the bottom of the worksheet.

15. Key **Prob 2-1** and press Enter.

16. Save your file as **Ch2**.

Histogram Data Analysis Feature

Problem 2-2. You can also use Excel®'s Histogram Data Analysis feature to create a histogram.

1. Retrieve the file **high temps**.

2. From the Ribbon, select Data tab. Select Data Analysis. Select Histogram. Click OK.

3. In the Input Range textbox key **A2:A51**

4. In the Bin Range textbox key **B2:B8**

The results can be shown on a separate sheet or displayed on the same sheet as your raw data. You will put the histogram on a separate sheet.

5. Select the radio button for New Worksheet Ply:

6. Select the check box for Chart Output. Click on OK.

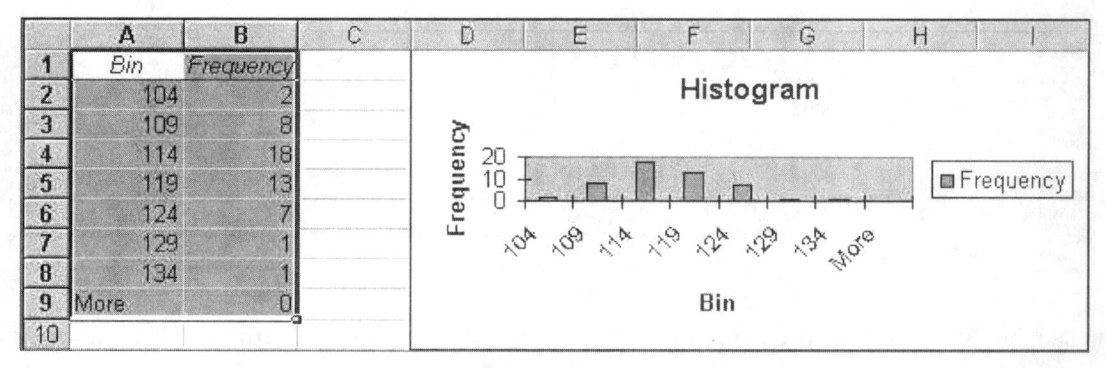

The frequency and histogram are displayed on a new sheet of your workbook. To make your chart look like a traditional histogram, do the following.

7. Click anywhere inside the chart. A frame will appear around the chart. The frame will have handles for resizing the chart. The handles appear as dots in the frame.

8. Put your mouse arrow on the bottom left handle of the frame. It will change to a line with an arrow at each end. Click, hold and drag the frame down to the bottom left corner of cell C14. This will enlarge the chart.

9. Double click on any one of the bars in the histogram. You will notice the Ribbon will change and a new set of tabs labeled Chart Tools will appear. The Chart Tools tabs are the Design, Layout and Format tabs.

10. Under the Layout tab, select Format Selection.

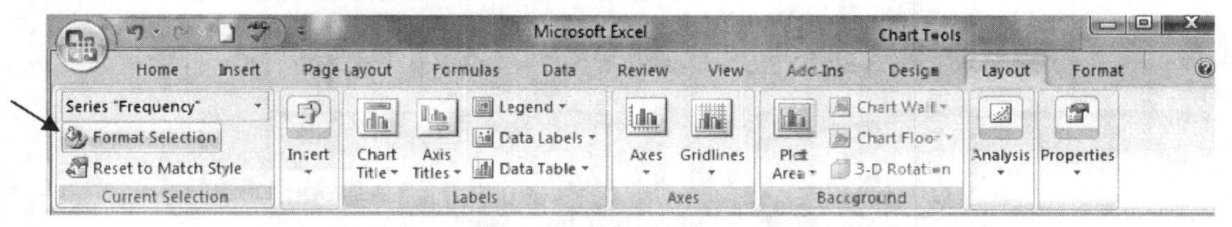

11. At the top of the Format Data Series window, select Series Options.

12. Click and drag the scroll bar in the Gap Width box to "No Gap". You may also type **0** in the text box. Click Close.

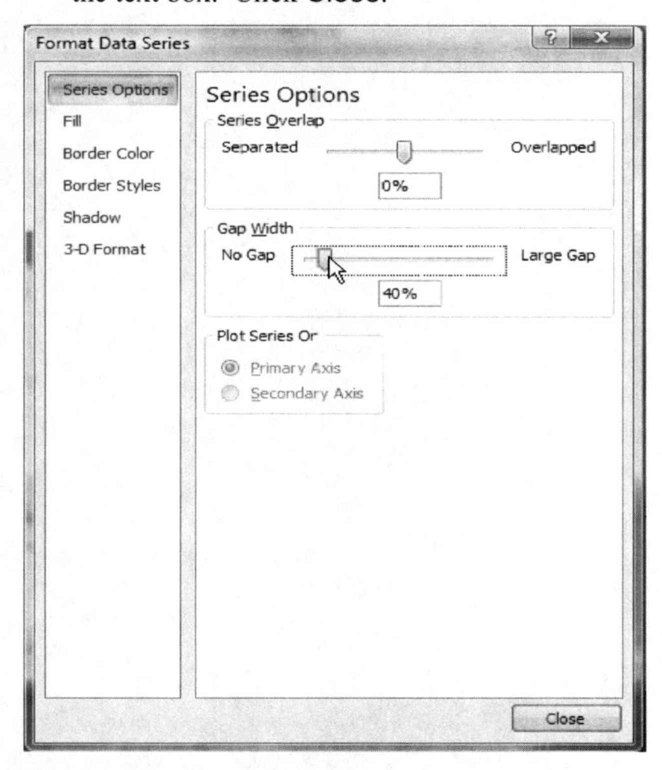

This displays your chart in the form of a histogram.

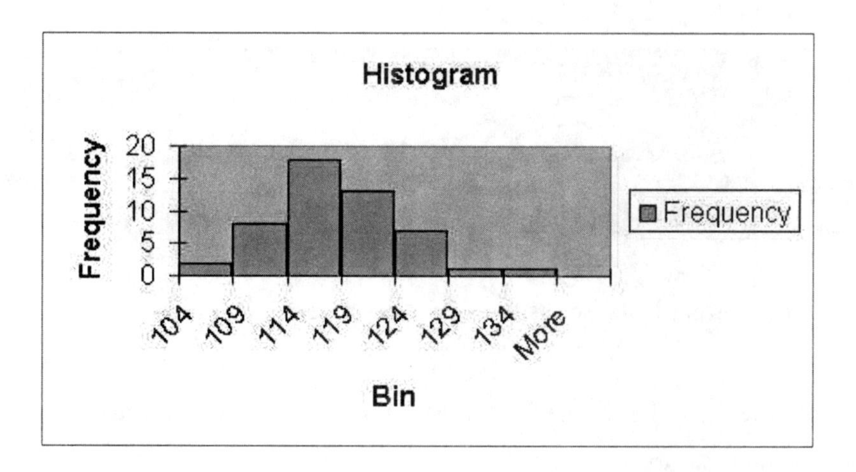

12. Save your file as **Ch2-prob2**

Printing Charts and Graphs

To print just the chart make sure the frame shows around the chart.

1. Click the Office Button
2. Select the Print

To print *both* the data and the chart, make sure the handles *do not* show on the box. (Click outside of the graph to make the handles disappear before printing.)

Using Excel® to Create Charts and Graphs

The following exercises will take you step by step through the creation of several charts.

Important Note: When creating charts and graphs key all related data in a single column. If labels extend over into the next column, either abbreviate to fit in one column or widen the column to accommodate the longest label.

Creating a histogram

Problem 2-3. You can use Excel® to construct your own histogram using the midpoints and frequency of a frequency distribution.

1. On a new worksheet, enter the data for the midpoint and frequency of the record high temperatures of the 50 states as shown.

	A	B
1	Temp	Frequency
2	102	2
3	107	8
4	112	18
5	117	13
6	122	7
7	127	1
8	132	1

2. Highlight cells A1:B8.

3. Select the Insert tab.

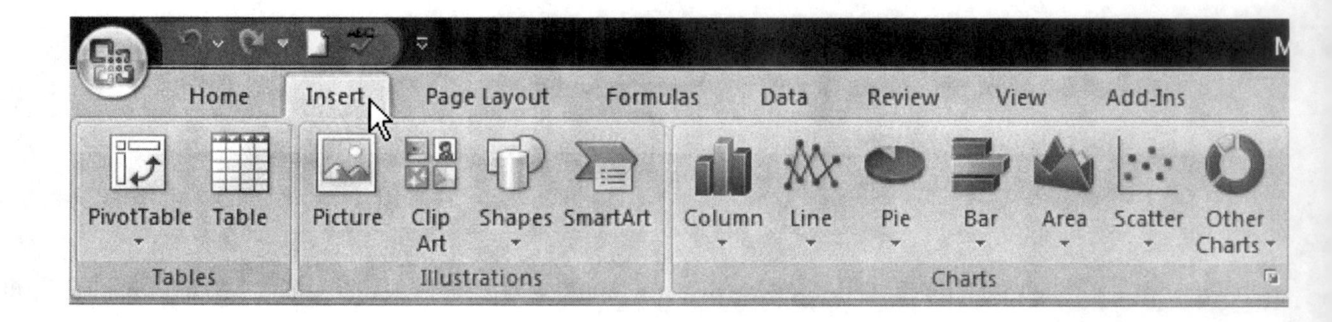

4. Click the Column drop down menu in the Charts group. Choose the upper left Column chart as shown below. (You can preview all available chart types by selecting one chart type then selecting All Chart Types on the drop down window.)

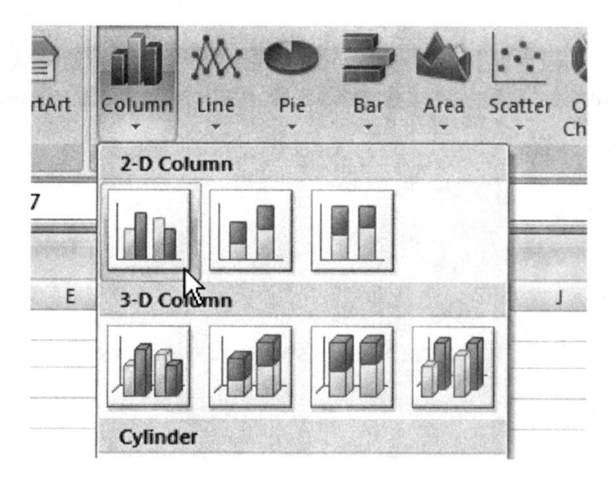

5. Click anywhere on the chart then click on the Design tab. Click the Select Data button. You will bring up the Select Data Source window.

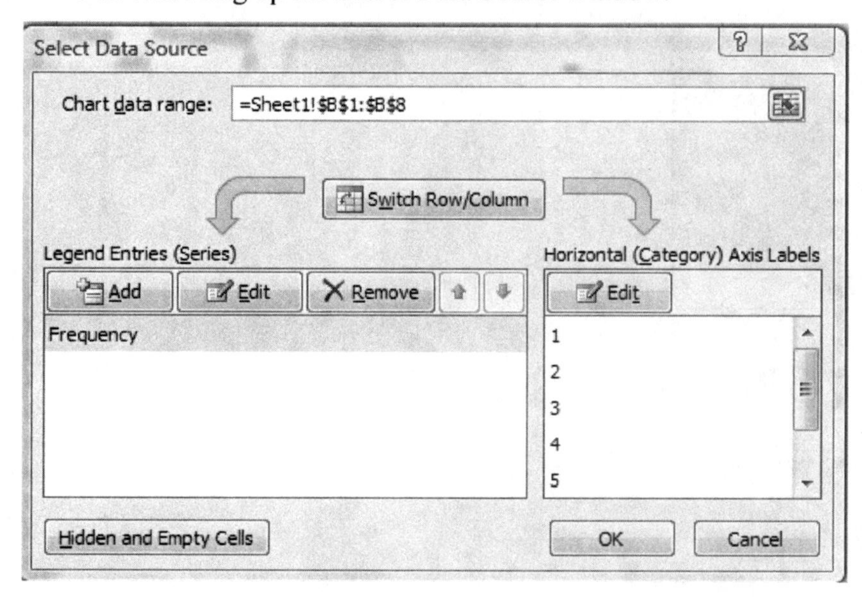

6. On the left under the Legend Entries (Series) list box, Temp should be highlighted. Select Remove. (We don't want the temperatures as part of the values).

7. On the right in the Horizontal (Category) Axis Labels textbox, click on the Edit at the far right.

8. Put your cursor on cell A2, click, hold and drag to cell A8. There will be a running box around cells A2:A8. Press <Enter>. This identifies the first column as the X-axis labels. Click Next.

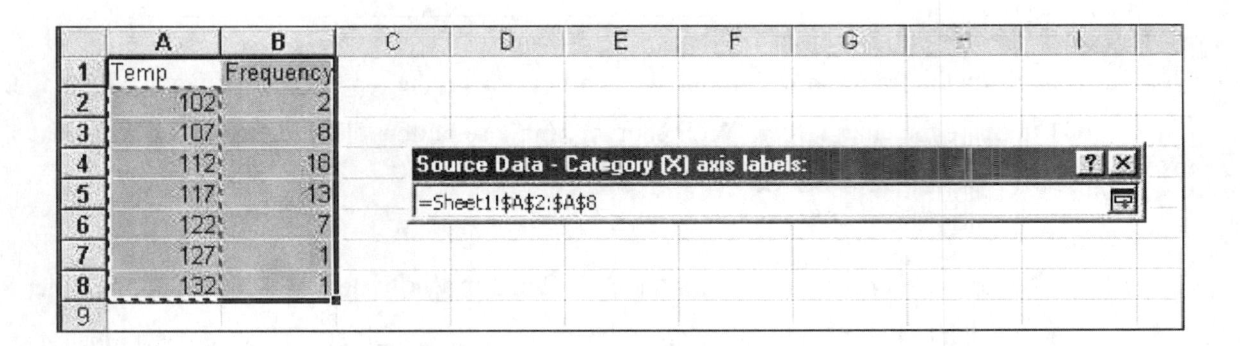

9. Click on the default chart title Frequency. Delete the default title, key **Record High Temperatures** in its place.

10. Select the Layout tab and click the Axis Titles button. Choose the following:
    a. Primary Horizontal Axis Title → Title Below Axis. Type **Temperature** to replace the default horizontal title.
    b. Primary Vertical Axis Title → Rotated Title. Type **Frequency** to replace the default vertical title.

11. Click the Legend button and Select None.

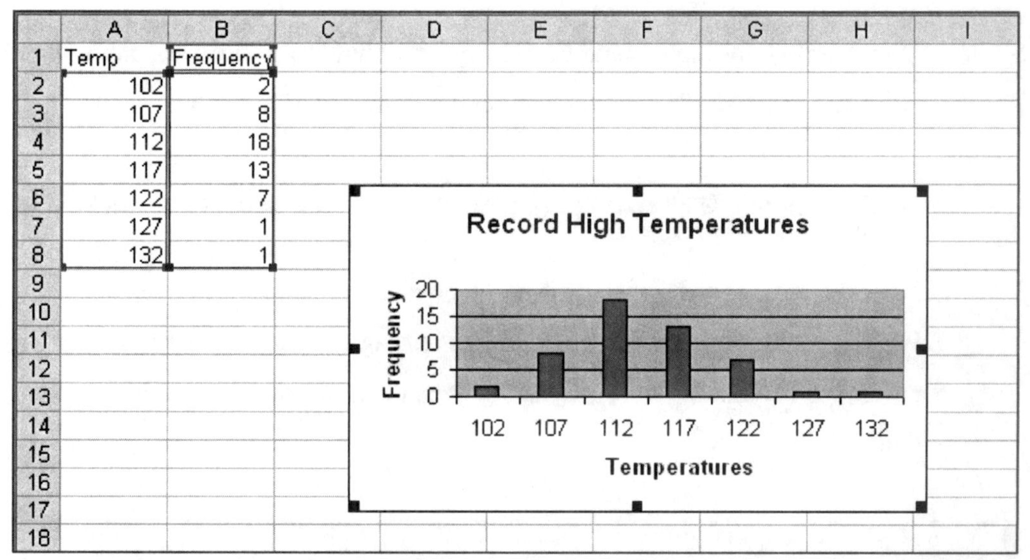

12. Make sure the frame shows around the chart. Click on the top left corner handle and drag to C1 to make the chart larger. With your **right** mouse button, click on one of the columns. Select Format Data Series.

13. At the top, select the Series Options tab. In the Gap Width box, click and hold the slider bar. Drag it to the left until the Gap width reads 0. Select Close. This step was illustrated in a previous example.

14. With your **right** mouse button, click on the Category Axis (Any of the numbers across the bottom of your chart along the X-axis). Select Font option.

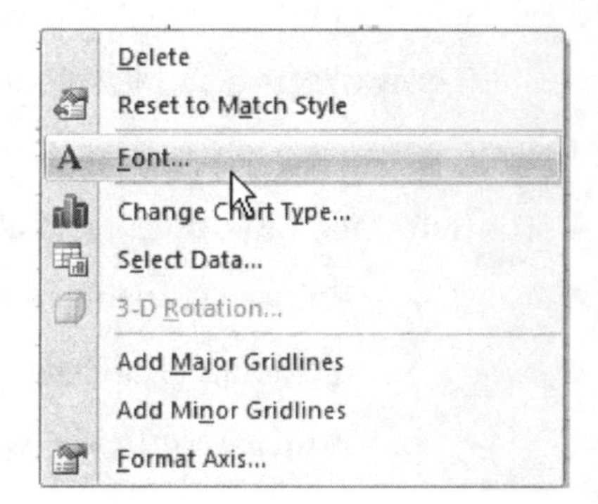

15. Under Size click on **8** (Click on the up arrow if 8 does not show). Select Close.

16. With your **right** mouse button, click on the Value Axis. (Any of the numbers along the side of your chart in the Y-axis.)

17. Select the Font option. Under Size click on **8**. Select OK.

This completes your histogram. You can also make changes to your Chart title and Axis titles using the same method.

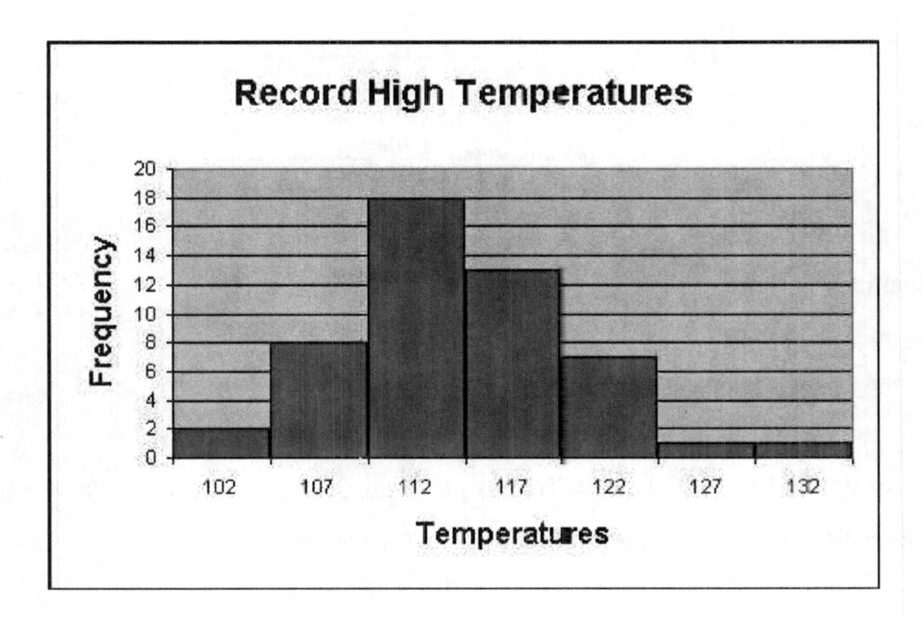

Save your workbook as **histogram**.

Compare this chart with the chart from Ch2-prob3 created using the histogram function. The charts are the same. The only difference is the X-axis across the bottom.

Pareto Charts

Problem 2-4. The table shown is the average cost per mile for passenger vehicles on state turnpikes. Construct a Pareto chart for the data.

State	Number
Indiana	2.9
Oklahoma	4.3
Florida	6.0
Maine	3.8
Pennsylvania	5.8

	A	B
1	State	Cost
2	Indiana	2.9
3	Oklahoma	4.3
4	Florida	6
5	Maine	3.8
6	Pennsylva	5.8
7		

1. Enter the data on a new worksheet as shown.

Notice you abbreviated Pennsylvania to fit in the cell. To make a Pareto chart, the data needs to be arranged in descending order from highest to smallest according to frequency.

2. Select cell B2.

3. Under the Data tab, select the Sort Descending icon. 

This sorts the data in descending order so you can create a Pareto chart. The same steps are used as creating a histogram.

	A	B
1	State	Cost
2	Florida	6
3	Pennsylva	5.8
4	Oklahoma	4.3
5	Maine	3.8
6	Indiana	2.9
7		

4. Highlight cells A2:B6.

5. Select the Insert tab on the Ribbon.

6. Click the Column drop down menu in the Charts group. Choose the upper left Column chart as in the histogram.

7. Select the Layout tab and click the Chart Title button. Choose Above Chart. Key **Average Cost Per Mile on State Turnpikes**.

8. Select the Layout tab and click the Axis Titles button. Choose the following:

a. Primary Horizontal Axis Title → Title Below Axis. Key **State** and Enter.

b. Primary Vertical Axis Title → Rotated Title. Key **Cost** and Enter.

9. Click the Legend button and Select None.

10. Click on the chart to ensure sure the frame shows around the chart. Click on the top left corner handle and drag to C1 to make the chart larger.

10. With your **right** mouse button, click on one of the columns. Select Format Data Series.

11. At the top of the window, select the Series Options tab. In the Gap Width box, click and hold the slider bar. Drag it to the left until the Gap Width reads 0. Select Close.

12. Change the font on both axes. **Right** click your mouse arrow on any number on an axis. Select Font. Under Size choose **8**. Click OK. Repeat for the second axis.

14. **Right** click your mouse arrow on the Chart Title **Average Cost Per Mile on State Turnpikes.** Select the Font. Change the Size to **8**. Select OK.

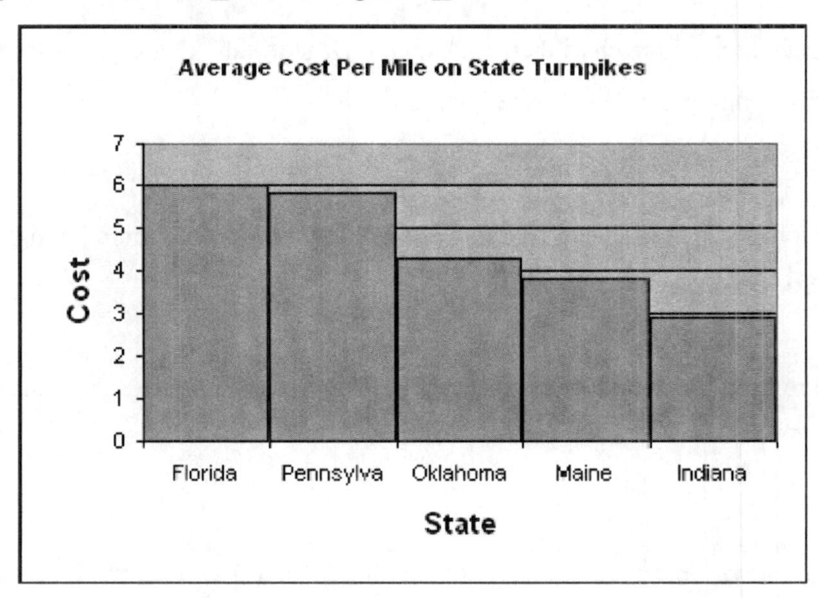

You may make any other changes to the chart if you wish. Save your file as **Pareto**.

Time Series Graph

Problem 2-5. The number (in millions) of vehicles, both passenger and commercial, that used the Pennsylvania Turnpike for the years 1999 through 2003 is shown. Draw a time series graph for the data.

Year	Number
1999	156.2
2000	160.1
2001	162.3
2002	172.8
2003	179.4

	A	B
1	Year	Number
2	1999	156.2
3	2000	160.1
4	2001	162.3
5	2002	172.8
6	2003	179.4
7		

1. On a new work sheet enter the data for the years 1999-2003 as shown.

2. Highlight cells A1:B6. Under the Insert tab, select Line chart style.

3. Select the left line chart in the middle row.

4. Click anywhere on the chart then click on the Design tab. Click the Select Data button. You will bring up the Select Data Source window.

5. On the left under the Legend Entries (Series) list box, Year should be highlighted. Select Remove.

6. On the right in the Horizontal (Category) Axis Labels textbox, click on the Edit at the far right.

7. Put your cursor on cell A2, click, hold and drag to cell A6. There will be a running box around cells A2:A6. Press <Enter>. This identifies the first column as the X-axis labels. Click Next.

8. Click on the default chart title Number. Delete the default title, key **Vehicles (in millions) Using Pennsylvania Turnpike** in its place.

9. Select the Layout tab and click the Axis Titles button. Choose the following:
    a. Primary Horizontal Axis Title → Title Below Axis. Type **Year** to replace the default horizontal title.
    b. Primary Vertical Axis Title → Rotated Title. Type **Vehicles (in millions)** to replace the default vertical title.

10. Click the Legend button and Select None.

11. Click the Gridlines tab. Select Primary Horizontal Gridlines then click None.

12. Make sure the frame shows around the chart. Click on one of the frame's corners and drag to enlarge the graph.

13. Change the font on numbers for both axes. **Right** click your mouse arrow on any number on an axis. Select Font. Under Size click on **8**. Click on Close. Repeat for the second axis.

14. **Right** click your mouse arrow on the Value Axis Title **Vehicles (in millions)**. Select Font. Under Size click on **10**. Select OK.

15. **Right** click your mouse arrow on the Category Axis Title **Year**. Select Font. Under Size click on **10**. Select OK.

Save your worksheet as **Time Series**. You can make several changes to your charts. Experiment. If you don't like the results, select the undo icon from the Tool bar.

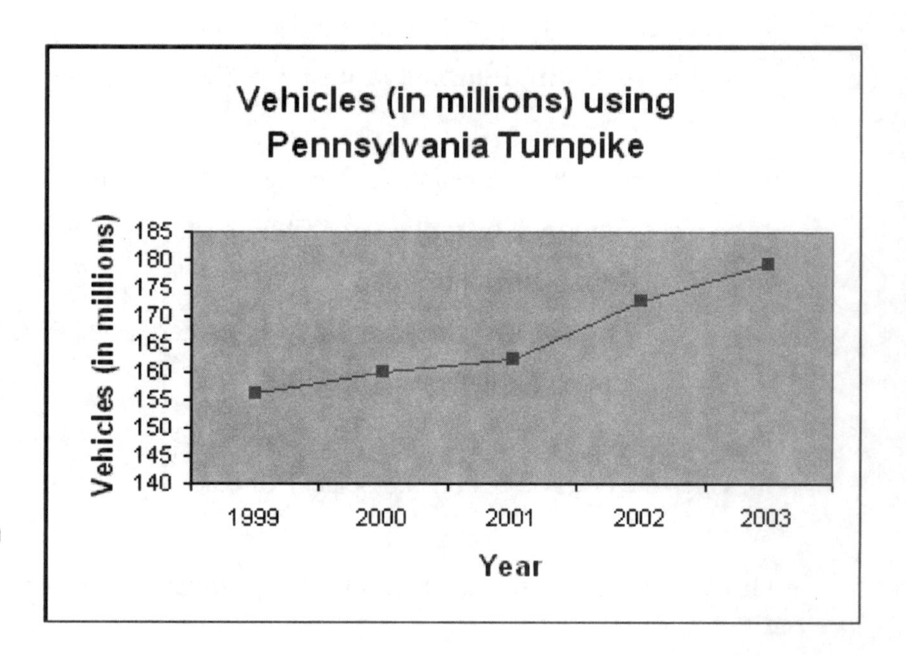

The Frequency Polygon

You can use the same method as creating a Time series graph to create a Frequency Polygon. Use the midpoints of your class limits as the X-axis and the frequency as your data on a line graph.

Pie Graphs

Problem 2-6. Use Excel® to create a pie chart.

1. Retrieve the file **Snacks** that you created in Chapter 1. Highlight cells A1:B6.
2. Under the insert tab, select Pie. Select the 1st pie chart (upper left chart).
3. In the Layout tab select Chart Title. Select the Above Chart option. Key **Popular Super Bowl Snacks.**
4. Still under the Layout tab select Data Labels. Select More Data Label Options.

On the Format Data Labels Window that appears you will use the Label Options, Number and Alignment menus.

5. Select the Label Options menu then do the following.

   a. Check Category Name and Percentage.

   b. Select the radio button for Inside End.

c. Check the Include legend key in label.

d. In the Separator box select (space).

e. Deselect any other options.

6. Select the Number menu.
   a. Choose Percentage.
   b. Key **1** under Decimal Place.

7. Select the Alignment menu and set the Custom Angle to 15°.
8. Place your mouse arrow on one of the snack labels in the chart. **Right** click your mouse. Select Font. Under Size click on **8**.

9. Make sure the frame shows around the chart. Click on one of the corners and drag to enlarge the chart.

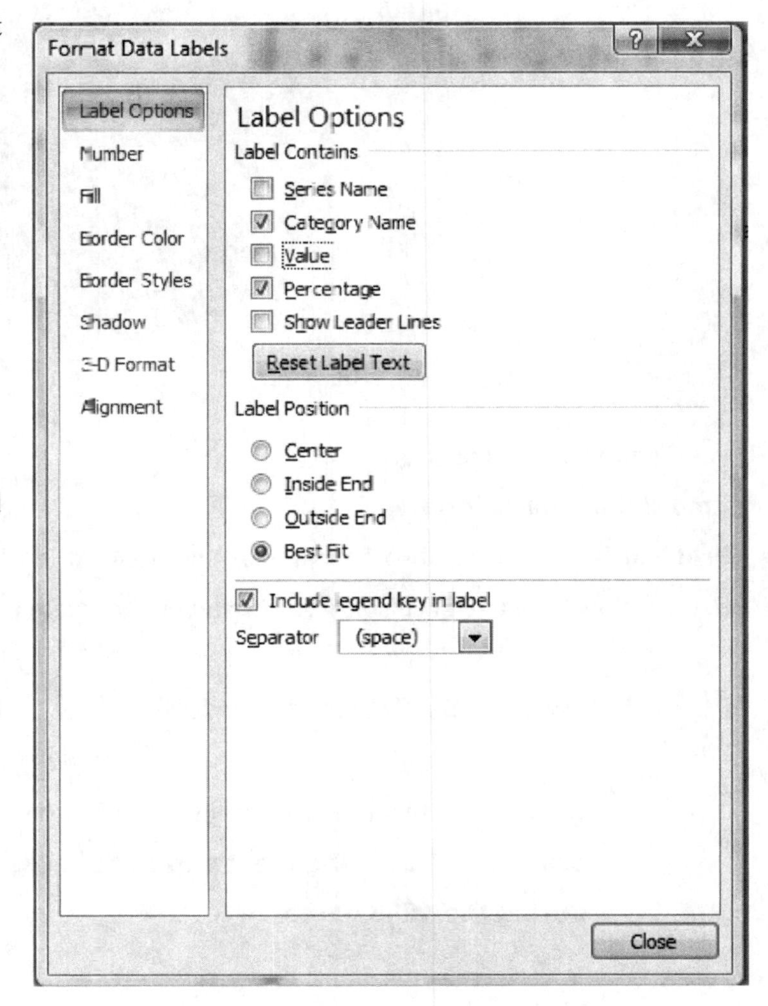

Click in a cell that is off the chart. Your chart will look similar to the one below.

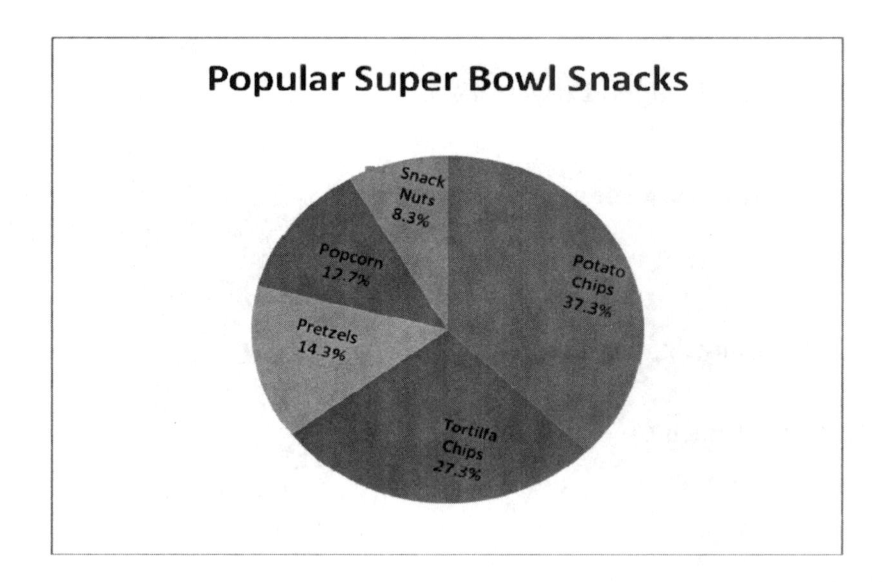

Save your worksheet as **pie.**

Embedding Charts in a Report

Problem 2-7. When creating a report describing data, it is often helpful to include a visual illustration. Excel® allows you to embed spreadsheets and/or charts into a Microsoft Word document.

1. Open Microsoft Word for Windows.

2. Prepare your report leaving room for your Excel® object. For this report key **The following graph illustrates the most popular snacks consumed by Super Bowl fans during the 1998 game.**

3. Minimize Word by clicking on the minimize box in the top right corner of your screen. It looks like a box with a small line at the bottom.

4. Open Microsoft Excel® if you are not already in it. Create the report or chart you want to use in your Word document. In this case you will retrieve a file you created earlier.

5. Click the Office Button. Select Open. Select your pie file . Click Open.

6. To insert a copy of the pie graph into the report, select the chart. Make sure that the frame appears. Under the Home tab select the Copy icon.

7. Minimize Excel® by clicking on the minimize box in the top right hand corner of your screen.

8. Activate Word again by clicking on the Microsoft Word button located on the task bar at the bottom of the screen, next to the start button.

9. Position your cursor at the point you wish to insert the pie chart into the report. From the Home tab. Select Paste Special.

10. Select Microsoft Excel® Chart Object. Select OK.

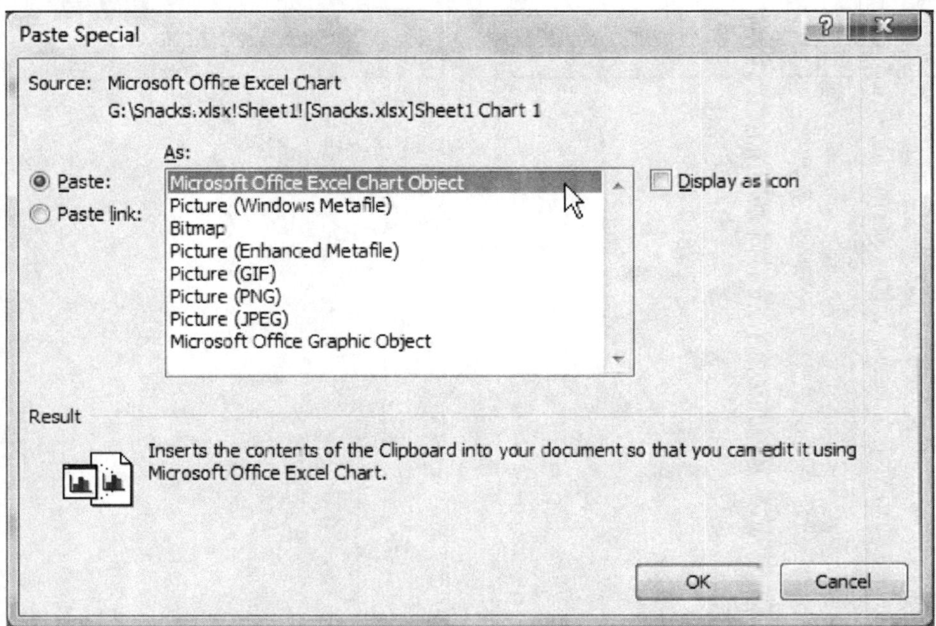

11. Place your mouse arrow on the chart in Word. You can move the chart by clicking and dragging, or you can resize the chart with the frame.

12. Click the Office Button and select Save As. Select the desired version of Word. In the File name text box, key **snack report**. Click Save.

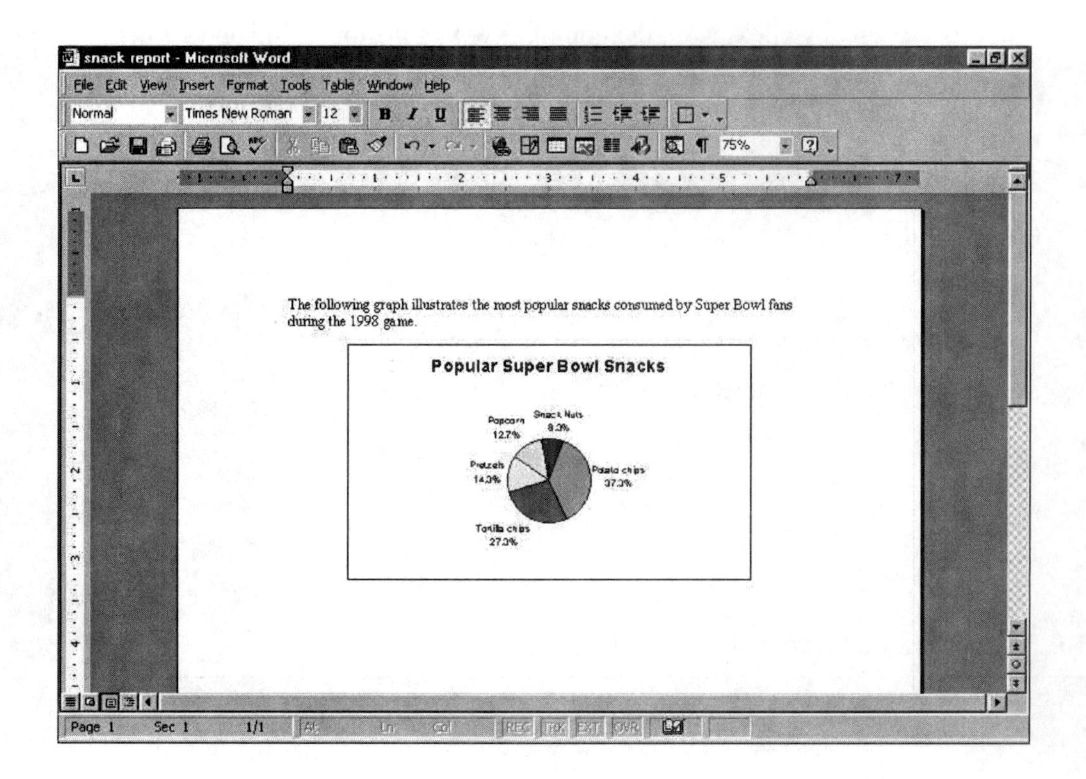
snack report - Microsoft Word
File Edit View Insert Format Tools Table Window Help
Normal
Times New Roman
12
75%
The following graph illustrates the most popular snacks consumed by Super Bowl fans during the 1998 game.
Popular Super Bowl Snacks
Popcorn 12.7%
Snack Nuts
Pretzels 14.0%
Potato chips 37.0%
Tortilla chips 27.0%
Page 1
Sec 1
1/1

Practice Exercises taken from textbook. (Key all related data in a single column.)

2-1. The heights in inches of commonly grown herbs are shown below. Construct a frequency distribution, using six classes. (Textbook Exercise 2-2, Problem 8)

18	20	18	18	24	10	15
12	20	36	14	20	18	24
18	16	16	20	7		

2-2. In a study of reaction times of dogs to a specific stimulus, an animal trainer obtained the following data, given in seconds. Construct a histogram and frequency polygon. Use the midpoints for both. (Textbook Exercise 2-3, Problem 8)

Class limits	Frequency
2.3 – 2.9	10
3.0 – 3.6	12
3.7 – 4.3	6
4.4 – 5.0	8
5.1 – 5.7	4
5.8 – 6.4	2

2-3. The World Roller Coaster Census Report lists the following number of roller coasters on each continent. Represent the data graphically, using a Pareto chart. (Textbook Exercise 2-4, Problem 4)

Africa	17
Asia	315
Australia	22
Europe	413
North America	643
South America	45

2-4. Draw a time series graph to represent the data for the number of airline departures (in millions) for the given years. (Textbook Exercise 2-4, Problem 6)

Year	1996	1997	1998	1999	2000	2001	2002
Number of departures	7.9	9.9	10.5	10.9	11.0	9.8	10.1

2-5. The following elements comprise the earth's crust, the outermost solid layer. Illustrate the composition of the earth's crust with a pie graph. (Textbook Exercise 2-4, Problem 12)

Oxygen	45.6%
Silicon	27.3
Aluminum	8.4
Iron	6.2
Calcium	4.7
Other	7.8

CHAPTER 3
DATA DESCRIPTION

Over 100 years ago H. G. Wells noted, "Statistical thinking will one day be as necessary for efficient citizenship as the ability to read and write." That day has arrived. Today, we cannot avoid being bombarded with all sorts of numerical data. Statistical techniques are used extensively in almost all career fields: social science, physical science, marketing, accounting, quality control, health science, education, professional sports, and politics to name just a few. This chapter will show how easy it is to use Excel® to find measures of central tendency and dispersion- measures that are essential to using and understanding statistical data.

Chapter 2 used Excel® to describe data using graphs and charts. In this chapter we will use Excel® to find a single value or an average to describe a set of data. This single value is referred to as a measure of central tendency. We often need a single number to represent a set of data – one number that can be thought of as being "typical" of all the data. Most people think of arithmetic mean when they hear the word average. However, there are several measures of central tendency including median, mode, weighted mean to name a few. This chapter will show how to:

- Use Excel®'s predefined formulas - called worksheet Functions - to perform calculations of arithmetic mean, median, mode, weighted mean, standard deviation, and variance.
- Use Excel®'s Descriptive Statistics Analysis ToolPak to find measures of central tendency and dispersion.

Using Worksheet Functions

Note: The structure of the function is critical in the performance of that function. Be careful to pay close attention to the use of the equals sign, the function name, opening and closing brackets, spaces between characters, and use of commas.

Finding the Mean

Problem 3-1: The data represent the number of days off per year for a sample of individuals selected from nine different countries. Find the mean.

20, 26, 40, 36, 23, 42, 35 24, 30

1. On a new worksheet, key **Mean** in A1, **Days off** in A3 and the numbers in A4:A12. In A13, key **Mean=**

2. In B13, key =**AVERAGE(A4:A12)**. Excel® uses *average* for the *arithmetic mean.*

3. From the Home tab, use the Decrease Decimal icon to round B13 to one decimal place.

The answer of 30.7 is displayed in cell B13.

4. Bold the contents of A1 and B13, using the icon on the Home tab.

Finding the Median

Problem 3-2: The number of tornadoes that have occurred in the United States over an 8-year period follows. (Remember; do not use commas when keying large numbers.)

Find the median: 684, 764, 656, 702, 856, 1133, 1132, 1303

1. On the same worksheet, key **Median** in C1, **Tornadoes** in C3 and the numbers in C4:C11. In C12, key **Median=**

2. In D12, key =**MEDIAN(C4:C11)**

The median 810 is displayed in cell D12. (Since there is an even number of data there is no single middle number. The median is the average of the middle *two* numbers, 764 and 856.)

3. Bold the contents of C1 and D12.

Finding the Mode

Problem 3-3: The following data represents the duration (in days) of U.S. space shuttle voyages for the years 1992-94.

Find the mode:	8,	9,	9,	14	8,	8,	10,	7,	6,
	9,	7,	8,	10	14,	11,	8,	14,	11

1. On the same worksheet, key **Mode** in E1, **Voyages** in E3, and the data listed above in E4:E21. In E22, key **Mode=**

2. In F22, key =**MODE(E4:E21)**

The mode 8 is displayed in cell F22. If there is *no* mode, Excel® will display #N/A in that cell. If there is *more than one* mode Excel® will display the one that occurs first in the string of data.

3. Bold the contents of E1 and F22.

Finding the Standard Deviation and Variance

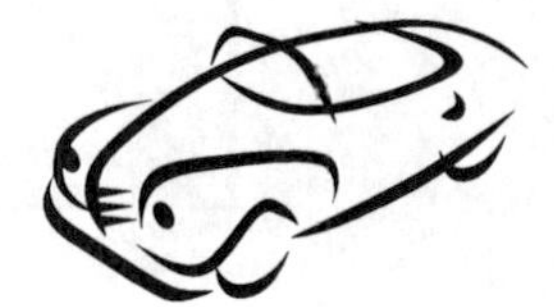

Problem 3-4: Find the sample variance and standard deviation for the amount of European auto sales for a sample of 6 years shown. The data are in millions of dollars.

11.2 11.9 12 12.8 13.4 14.3

1. On the same worksheet, key **Standard Deviation and Variance** in G1, **Auto Sales** in G3 and the data in G4:G9. In G10, key **StdDev=**, in G11 key, **Variance=**

2. In H10, key =**STDEV(G1:G9)**

3. In H11, key =**VAR(G1:G9)**

5. From the Tool bar, use the Decrease Decimal icon to round H10 and H11 to two decimal places.

4. 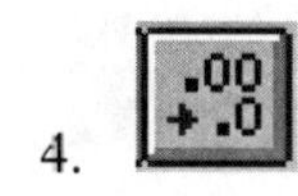

The standard deviation of 1.13 is displayed in H10 and the variance of 1.28 is displayed in cell H11.

6. Bold the contents of G1, H10, and H11.

Your final worksheet should look as follows:

	A	B	C	D	E	F	G	H	I	J
1	**Mean**		**Median**		**Mode**		**Standard Deviation and Variance**			
2										
3	Days off		Tornadoes		Voyages		Auto Sales			
4	20		684		8		11.2			
5	26		764		9		11.9			
6	40		656		9		12			
7	36		702		14		12.8			
8	23		856		8		13.4			
9	42		1133		8		14.3			
10	35		1132		10		StdDev=	**1.13**		
11	24		1303		7		Variance=	**1.28**		
12	30		Median=	**810**	6					
13	Mean=	**30.7**			9					
14					7					
15					8					
16					10					
17					14					
18					11					
19					8					
20					14					
21					11					
22					Mode=	**8**				
23										

Save your file as **Ch3-prob 1-4**

All of the functions that you have used in this chapter so far are also accessible by using the Insert Function icon located on the formula bar. You will learn more about the use of the Insert Function in Chapter 4.

Using Excel®'s Descriptive Statistics Analysis ToolPak

Excel®'s Descriptive Statistics Analysis ToolPak allows you to find the mean, standard error, median, mode, standard deviation, sample variance, kurtosis, skewness, range, minimum, maximum, sum, count, largest number by position, smallest number by position, and the level of confidence simply by entering the numerical data. If Analysis Tools doesn't appear in the Add-Ins available list box, you may need to add the Analysis ToolPak using a custom installation of the Microsoft Excel® Setup program. The following problem will demonstrate how to use Excel® to compute measures of central tendency and dispersion.

Problem 3-5: The following data represent the number of listeners (in thousands) of 15 radio stations in the 6:00 to 9:00 A.M. time slot in Pittsburg. Find the descriptive statistics. (Taken from Review Exercise 1 for demonstration purposes).

229 182 129 112 122 93 97 114 95 114 60 89
75 70 68

1. On a new worksheet, key your data in column A.
2. Under the Data tab select Data Analysis. Choose Descriptive Statistics. Click on OK.

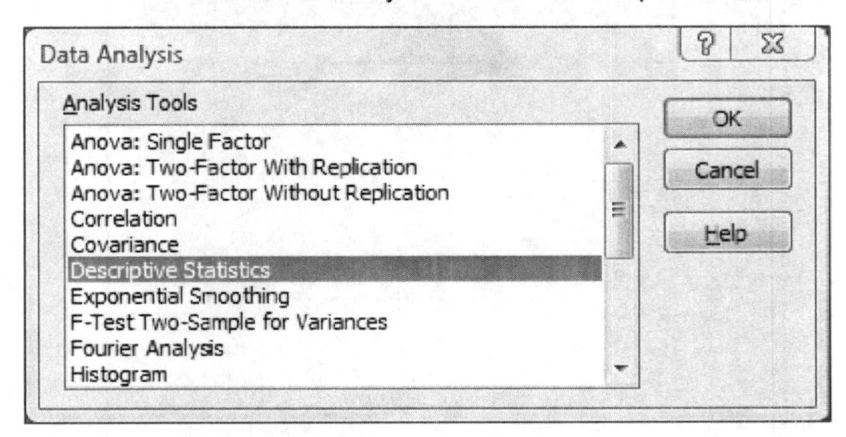

Remember: The Analysis ToolPak option must be installed and selected before proceeding as noted in the Chapter 2 instructions.

3. For Input Range, key **A1:A15** or you may click your mouse on cell A1 then hold and drag to highlight cells A1:A15. Your input range will vary depending on the number in your sample.

4. Grouped by Columns, should be selected.

5. Labels in First Row should not be selected unless the column contains a label.

6. Select Output Range. Select the text box, key in **C1**.

(New Worksheet Ply and New Workbook should not be selected.)

7. Select Summary statistics.

8. Select Confidence Level for Mean. Select the text box, key in **90** %.

9. Select Kth Largest. Select the text box, key in **2**.

10. Select Kth Smallest. Select the text box, key in **2**. Click OK.

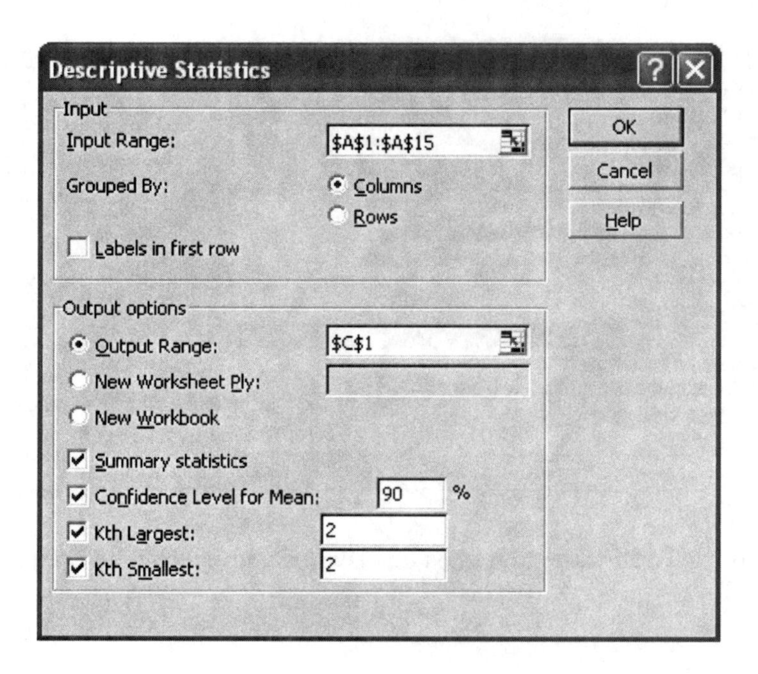

The descriptive statistics have been computed automatically.

	A	B	C	D
1	229		Column1	
2	182			
3	129		Mean	109.9333
4	112		Standard E	11.56739
5	122		Median	97
6	93		Mode	114
7	97		Standard D	44.8003
8	114		Sample Va	2007.067
9	95		Kurtosis	2.82704
10	114		Skewness	1.601328
11	60		Range	169
12	89		Minimum	60
13	75		Maximum	229
14	70		Sum	1649
15	68		Count	15
16			Largest(2)	182
17			Smallest(2	68
18			Confidence	20.37375
19				

HINT: After you get your results, cross-check the *count* in the output table to make sure it contains the correct number of sample items.

To make your work easier to read do the following:

1. Label the data column
    a. Right mouse click on Cell A1. Select Insert.
    b. Select Shift cells down. Click OK.
    c. Key **Number of Listeners** in the now empty cell A1.

2. Place your mouse arrow in the column heading between columns C and D. The arrow will change to a thick black plus sign. Double click your left mouse button. Column C will automatically widen to accommodate the longest description.

2. Highlight cell D3. From the Home tab, place your mouse pointer on Decrease Decimal icon. Click your left mouse button. Round the Mean to one decimal place. Repeat this action for the Standard deviation in cell D7 and it is easier to read.

3. Format cells D4, D9, D10, and D18 to all have three decimal places.

The chart is now easier to read.

	A	B	C	D
1	Number of Listeners		*Column1*	
2	229			
3	182		Mean	109.9
4	129		Standard Error	11.567
5	112		Median	97
6	122		Mode	114
7	93		Standard Deviation	44.8
8	97		Sample Variance	2007.067
9	114		Kurtosis	2.827
10	95		Skewness	1.601
11	114		Range	169
12	60		Minimum	60
13	75		Maximum	229
14	70		Sum	1649
15	68		Count	15
16	89		Largest(2)	182
17			Smallest(2)	68
18			Confidence Level(95.0%)	24.810
19				

Save your file as **Ch3-prob5**

Excel®'s Descriptive Statistics Output Table contains all the measures of central tendency: mean, median and mode, the standard deviation and the variance you computed by using Excel®'s functions, plus others. You may not need the other computations but all are described as follows.

The **Mean** (109.9) is computed by dividing the sum (1649) by the count (15).

The **Median** (97) is a measurement of position in a ranked set of data. It is the middle number in a data set with an odd number of values. In an even set of numbers, it is the value halfway between the two middle values.

The **Mode** (114) is a measurement of frequency, it is the most frequently occurring value. When there are two or more values that appear the same number of times (duplicate modes), Excel® reports the value that appears first in the data set. In the data sets in which each value is unique, Excel® reports "#N/A."

The output table contains several measures of variation. The **Range** (169) equals the **Maximum** value (229) minus the **Minimum** value (60). Remember, with some data sets the range can be a misleading measure of variation since it only contains the two most extreme values.

The **Standard Deviation** (44.8) is the most common measure of variation or dispersion. In a normal or symmetrical set of data about 68 percent will be within plus or minus one standard deviation of the mean, 95 percent will be within plus or minus two standard deviations of the mean , and almost all of the data (99.7 percent) will be within plus or minus three standard deviations of the mean.

The **Variance** is the standard deviation squared. Excel®'s output table shows the sample standard deviation and variance computed using *n-1* in the denominator. To find the *population* standard deviation and the *population* variance, computed by using *n* as the denominator, use the STDEVP and VARP functions.

The **Largest (2)** and the **Smallest (2)** values in Excel®'s output table are the second largest (182) and the second smallest (68) number. These values can be used to eliminate outliers. They can also be used to estimate quartiles in data with a large number of frequencies. For example if you had 1600 in your data set you would divide the count (1600) by 4 and enter 400 as the largest and the smallest. The output table would then show the approximate third and first quartile.

The **Standard Error** (11.567) is the standard deviation divided by the square root of the sample size. It is a measure of uncertainty about the mean, and is used for statistical inference (confidence intervals, regression belts, and hypothesis tests.)

The **Confidence Level (90.0%)** (20.374) is half of the 90% confidence interval for the mean.

Kurtosis (2.827) measures the degree of peakedness in symmetric distributions. If a symmetric distribution is more peaked than the normal distribution, that is, if there are fewer values in the tails, the kurtosis measure is negative. If the distribution is flatter than the normal distribution, that is if there are more values in the tails than a corresponding normal distribution, the kurtosis measure is positive.

Skewness (1.601) is a measurement of the lack of symmetry in a distribution. If there are a few extreme small values and the tail of the distribution runs off to the left we say the distribution is negatively skewed and our skewness value would be negative. If there are a few extreme large values and the tail of the distribution runs off to the right, we say the distribution is positively skewed and the skewness value would be positive.

Finding the Weighted Mean

Excel® does not have a function for weighted mean so you will create a template.

Problem 3-6. A student received an A in English Composition I (3 credits), a C in Introduction to Psychology (3 credits), a B in Biology I (4 credits), and a D in Physical Education (2 credits). Assuming A = 4 grade points, B = 3 grade points, C = 2 grade points, D = 1 grade point, and F = 0 grade points, Find the student's grade point average.

1. On a new worksheet in A1, key **Weighted Mean.**

2. Key **Subject** in A3, **Credit** in B3, **Grade** in C3, **Product** in D3, **Weighted mean=** in A10

3. Key in the subjects **English, Psychology, Biology, Phys. Ed** in A4:A7 respectively.

4. Key the credits **3**, **3**, **4**, and **2** in B4:B7 respectively.

5. Key the grade points **4**, **2**, **3**, and **1** in C4:C7 respectively.

6. In D4, key =**A4*B4**.

7. Make D4 your active cell. Place your cursor on the bottom right handle. You will have a thick black plus sign. Click and drag to D5:D7.

8. Highlight B4:B7. From the Home tab choose the Sum icon.

9. Highlight D4:D7. From the Home tab choose the Sum icon.

10. In B10, key =**D8/B8**

11. Place your mouse arrow in the column heading between columns A and B. The arrow will change to a thick black plus sign. Double click your left mouse button Column A will automatically widen to accommodate the longest description.

12. Decrease the decimal of B10 to one decimal place.

13. Bold the contents of A1 and B10.

The grade point average is 2.7. Your output will look as follows.

	A	B	C	D
1	**Weighted Mean**			
2				
3	Subject	Credit	Grade	Product
4	English	3	4	12
5	Psychology	3	2	6
6	Biology	4	3	12
7	Phys. Ed.	2	1	2
8		12		32
9				
10	Weighted Mean =	**2.7**		
11				
12				

Save your file under the filename **Ch3-prob6**

Practice Exercises taken from textbook. (Key all related data in a single column)

Find the (a) mean, (b) median and (c) mode for Problems 3-1 and 3-2 using Excel®'s functions of AVERAGE, MEDIAN, and MODE.

3-1. The following data are the number of burglaries reported for a specific year for nine western Pennsylvania universities. (Textbook Exercise 3-2, Problem 3)

61, 11, 1, 3, 2, 30, 18, 3, 7

3-2. Twelve major earthquakes had Richter magnitudes shown here. (Textbook Exercise 3-2, Problem 7)

7.0, 6.2, 7.7, 8.0, 6.4, 6.2, 7.2, 5.4, 6.4, 6.5, 7.2, 5.4

3-3. The following data represent the area in square miles of major islands in the Caribbean Sea and the Mediterranean Sea. (Textbook Chapter 3, Review Exercise #2)

Use Excel®'s Data Analysis ToolPak to obtain the mean, median, mode, range, variance and standard deviation of each of the two following sets of data. Be sure to put all related data in a single column.

Caribbean Sea			Mediterranean Sea	
108	926	436	1927	1411
75	100	3339	229	95
5382	171	116	3189	540
2300	290	1864	3572	9301
166	687	59	86	9926
42804	4244	134		
29389				

3-4. Find the weighted mean price of three models of automobiles sold. The number and price of each model sold are shown in the following list. (Textbook Exercise 3-2, Problem 26)

Model	Number	Price
A	8	$10000
B	10	12000
C	12	8000

3-5. An investor calculated the following percentages of each of three stock investments with payoffs as shown. Find the average payoff. Use the weighted mean. (Textbook Chapter 3, Review Exercise # 8)

Stock	Percent	Payoff
A	30%	$10000
B	50%	3000
C	20%	1000

CHAPTER 4

PROBABILITY AND COUNTING RULES

Many important decisions are made on the probability of an event happening. In Chapter 4 you are introduced to the concept of probabilities. Excel® assists you in finding probabilities, and developing frequency distributions that can be illustrated using a graph or chart. You will learn to:

- Demonstrate the concept of classical probability by simulating a coin flip, both a small number and a large number of times. Then using the coin flip outcomes, develop a frequency distribution to illustrate Empirical probability concepts.
- Illustrate the law of large numbers comparing the means and standard deviations of a frequency distribution.
- Define and construct a sampling distribution of sample means.

In this chapter, you will also be introduced to the Insert Function in calculating the counting rules: permutations and combinations. As you go through this chapter you will:

- Use Excel®'s Insert Function to find the number of permutations in a subset.
- Use Excel®'s Insert Function to find the number of combinations in a subset.

Probability

Example 4-1. Using Excel® to simulate a coin flip.

1. On a new worksheet, enter the data as shown below.

	A	B	C	D	E	F
1	0=tails	1=heads				
2	Flip 1					Flip1
3	Flip 2					Flip 2
4						

2. Highlight A2:A3. Drag the lower right handle of A3 to A5.

3. Highlight F2:F3. Drag the lower right handle of F3 to F16.

4. In cell B2, key **=RANDBETWEEN(0,1)**

This gives you a random number of 0 (tails) or 1 (heads).

5. Make B2 your active cell. Drag the lower right handle of B2 to B5.

6. With cells B2 to B5 highlighted, select the AutoSum icon under the Home tab.

7. Cells B2 to B6 should now be highlighted. Drag the lower right handle of B6 to C6:D6.

8. In cell G2 key =**RANDBETWEEN(0,1)**

9. Make G2 your active cell. Drag the lower right handle of G2 to G16.

10. With Cells G2 to G16 highlighted, select the AutoSum icon under the Home tab.

11. Cells G2 to G17 should be highlighted. Drag the lower right handle of G17 to H17:I17.

You will notice as you make changes that worksheet changes. This represents new flips of the coin.

F9

12. Press the F9 key several times. Each time represents a different set of coin flips.

    You will notice with the set of 4 coin flips you often get all heads (4 - 1's) or no heads (no - 1's). But with the set of 15 coin flips you very seldom get below 3 or above 12 heads (1's).

	A	B	C	D	E	F	G	H	I	J
1	0=tails	1=heads								
2	Flip 1	1	1	0		Flip1	0	1	0	
3	Flip 2	1	0	0		Flip 2	1	1	0	
4	Flip 3	1	1	0		Flip3	1	0	0	
5	Flip 4	1	1	0		Flip4	1	1	1	
6		4	3	0		Flip5	0	0	0	
7						Flip6	0	0	0	
8						Flip7	1	0	1	
9						Flip8	0	1	1	
10						Flip9	0	0	0	
11						Flip10	0	0	0	
12						Flip11	1	0	0	
13						Flip12	0	0	1	
14						Flip13	1	0	0	
15						Flip14	1	0	1	
16						Flip15	1	0	0	
17							8	4	5	
18										

If you wish save your file as **Ch5-examp1**

Sampling From a Normal Population. The following example will further illustrate the law of large numbers.

Example 4-2. Birth weight of a newborn is a major concern for all new parents. Nationally, the 50-percentile birth weight of children born at full term (40 weeks) is 7.04 pounds. That is, the average or "normal" birth weight of a full term baby is 7.04 pounds. From a random sample of 237 full term babies born at Community Hospital, their mean weight was 7.04 with a standard deviation of .42. You will generate a random distribution then use samples from that distribution to illustrate the law of large numbers.

a. Generating a Random Distribution

1. On a new worksheet enter the data as shown.

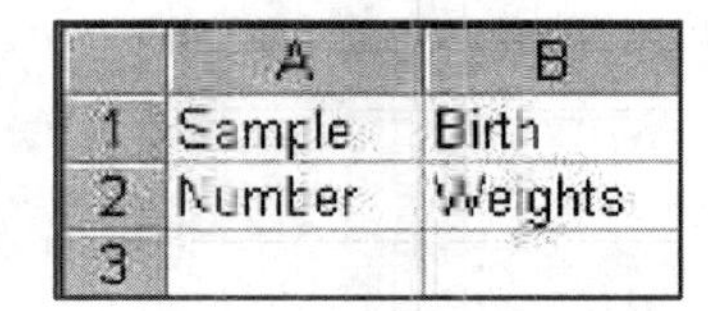

	A	B
1	Sample	Birth
2	Number	Weights
3		

2. In A3, key **1**. Press the <Enter> key.

3. Make A3 your active cell. From the Home tab, in the Editing group select the drop down arrow for the Fill icon. Select Series.

4. From the Series dialog box, select Series in Columns. Select Linear under Type. In the Step Value text box, key **1**. In the Stop Value text box, key **237**. Click OK.

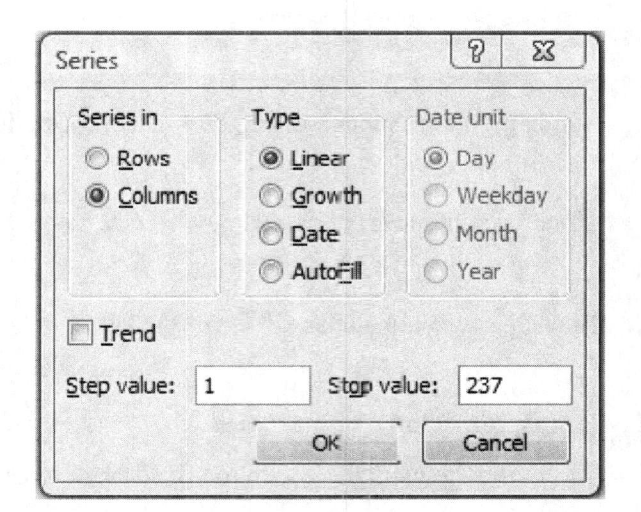

This enters the number 1-237 in A3:A239.

5. In B3, key **=NORMINV(RAND(),7.04,.42)**. Press <Enter>. (To create (), use the left and right parentheses)

This generates random numbers with a mean of 7.04 and a standard deviation of .42.

6. Make B3 your active cell. Place your mouse arrow on the lower right handle of B3. Make sure you have a thick, black plus sign. Click your left mouse button twice, rapidly.

This automatically fills the cells in B4:B239 with random numbers. Press the F9 (re-calculate) function key on your keyboard. Notice how the random birth weights change each time. Each time you perform a new procedure, the random values change. You will need a fixed set of random values for sampling. The values on your worksheet will be different from the examples that follow.

7. Highlight A1:B239. As you drag your mouse pointer below the worksheet it will highlight the lower rows. From the Home tab, select the Copy icon.

8. Make a new worksheet active. If there are no additional worksheets in your workbook click the Insert Worksheet tab located at the bottom of the Workbook.

9. On the new worksheet make A1 your active cell. From the Home tab, select the Paste drop down arrow and select Paste Special. You can also right mouse click on cell A1 to select Paste Special.

10. In the Paste Special dialog box, select the radio button by Values and click OK.

Your random numbers should now stay as a population from which to take samples.

If you wish, save your file as **Ch4-examp2-a**

b. Sample Sizes and Comparisons

1. Using the worksheet with the fixed random values, enter the contents in E1:H2 as shown below.

	A	B	C	D	E	F	G	H
1	Sample	Birth			Sample	Sample	Sample	Sample
2	Number	Weights			of 3	of 3	of 30	of 30
3	1	7.592116						
4	2	6.160641						
5	3	6.925447						

Remember: The Analysis ToolPak option must be installed and selected before proceeding as noted Chapter 2 instructions.

2. From the Data tab, select Data Analysis. Select Sampling. Click OK.

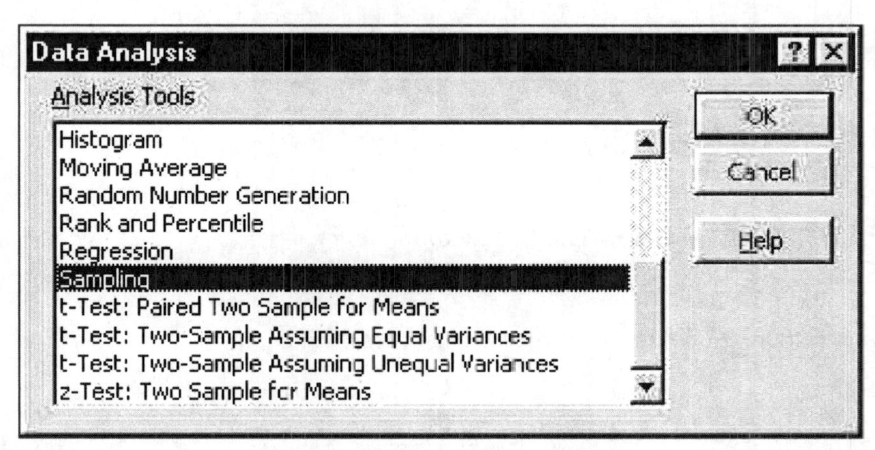

The Sampling dialog box appears.

3. Your cursor should be in the text box for Input Range. Key **B3:B239**.

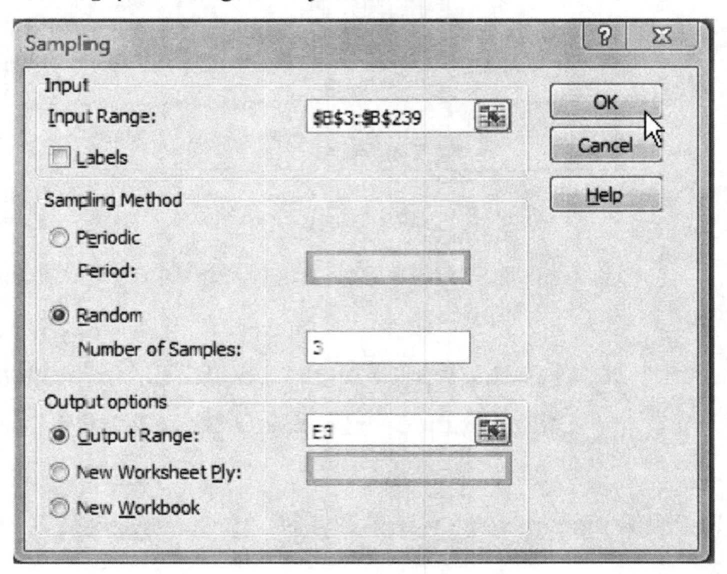

4. Select the radio button for Random. In the Number of Samples text box, key **3**. Select the radio button for Output Range. In the Output Range text box, key **E3**. Click OK.

From the random weights in B3:B239, 3 were selected.

5. Under the Data tab, select Data Analysis. Select Sampling. Click OK.

5. Leave the Input Range as **B3:B239**. Leave the Number of Samples as **3**. In the text box for Output Range, replace the contents with **F3.** Click OK.

From the random weights in B3:B239, 3 more weights were selected.

6. In D6:D7, key **Mean** and **StdDev**, respectively.

7. In E6, key =**Average(E3:E5)**

8. In E7, key =**STDEV(E3:E5)**

9. Highlight E6:E7. Drag the lower right handle of E7 to F7.

This computes the mean and standard deviation of the two samples of 3.

	A	B	C	D	E	F	G	H
1	Sample	Birth			Sample	Sample	Sample	Sample
2	Number	Weights			of 3	of 3	of 30	of 30
3	1	7.592116			6.830568	6.706177		
4	2	6.160641			7.599144	6.751228		
5	3	6.925447			7.108211	6.96647		
6	4	6.672467		Mean	7.179308	6.807958		
7	5	7.283647		StdDev	0.38919	0.139111		
8	6	6.818563						
9	7	7.477279						
10	8	7.409539						

10. Under the Data tab, select Data Analysis. Select Sampling. Click OK.

11. Leave the Input Range as B3:B239. In the Number of Samples text box, key **30.** In the Output Range text box, key **G3**. Click OK.

From the random weights in B3:B239, 30 were selected.

12. Under the Data tab, select Data Analysis. Select Sampling. Click OK.

13. Leave the Input Range as B3:B239. Leave the Number of Samples as **30**. In the Output Range text box, key **H3**. Click OK.

From the random weights in B3:B239, 30 more were selected.

14. In F33:F34, key **Mean** and **StdDev**, respectively.

15. In G33, key =**AVERAGE(G3:G32)**

16. In G34, key =**STDEV(G3:G32)**

17. Highlight G33:G34. Drag the lower right handle of G34 to H34.

This computes the mean and standard deviation of the two samples of 30.

18. Highlight A9:A30. (Column A will stop at Sample Number 28). From the Home tab, select Format. Select Hide & Unhide. Select Hide Rows.

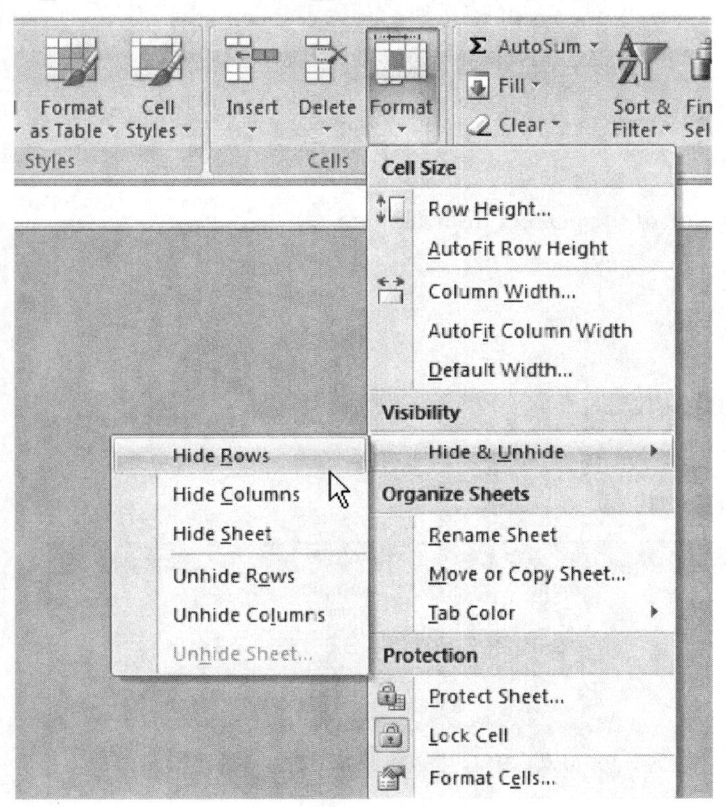

This hides rows 9-30 and brings your mean and standard deviation of all 4 samples closer together so you can compare them.

	A	B	C	D	E	F	G	H
1	Sample	Birth			Sample	Sample	Sample	Sample
2	Number	Weights			of 3	of 3	of 30	of 30
3	1	7.592116			6.830568	6.706177	6.65156	7.631414
4	2	6.160641			7.599144	6.751228	7.905913	6.710266
5	3	6.925447			7.108211	6.96647	6.809353	6.925447
6	4	6.672467		Mean	7.179308	6.807958	7.077121	7.486474
7	5	7.283647		StdDev	0.38919	0.139111	7.066425	7.087193
8	6	6.818563					7.063442	8.192396
31	29	6.929912					7.085614	7.704875
32	30	6.914709					7.084522	7.293902
33	31	6.233269				Mean	7.055496	6.962239
34	32	6.505549				StdDev	0.345667	0.408372
35	33	5.879552						

19. In C36:C37, key **% of diff mean** and **% of diff StdDev**, respectively.

20. In H36:H37, key **% of diff mean** and **% of diff StdDev**, respectively.

21. In E36, key **=ABS(E6-F6)/E6.**

This computes the absolute (positive) value of the percent of difference between the two means.

22. In E37, key **=ABS(E7-F7)/E7**

23. In G36, key **=ABS(G33-H33)/G33**

24. In G37, key **=ABS(G34-H34)/G34**

	A	B	C	D	E	F	G	H	I
1	Sample	Birth			Sample	Sample	Sample	Sample	
2	Number	Weights			of 3	of 3	of 30	of 30	
3	1	7.592116			6.830568	6.706177	6.65156	7.631414	
4	2	6.160641			7.599144	6.751228	7.905913	6.710266	
5	3	6.925447			7.108211	6.96647	6.809353	6.925447	
6	4	6.672467		Mean	7.179308	6.807958	7.077121	7.486474	
7	5	7.283647		StdDev	0.38919	0.139111	7.066425	7.087193	
8	6	6.818563					7.063442	8.192396	
31	29	6.929912					7.085614	7.704875	
32	30	6.914709					7.084522	7.293902	
33	31	6.233269				Mean	7.055496	6.962239	
34	32	6.505549				StdDev	0.345667	0.408372	
35	33	5.879552							
36	34	6.865114	% of dif mean		0.051725		0.013218	% of dif mean	
37	35	6.706177	%of diff StdDev		0.642562		0.181404	%of diff StdDev	
38	36	7.169123							

This allows you to compare the means and standard deviations.

25. Highlight E36:G37. From the Home tab, select the Percent Style icon. 

26. From the Home tab, select the Increase Decimal icon. Click one time. 

This changes the format of the differences to percents with one place past the decimal.

	A	B	C	D	E	F	G	H	I
1	Sample	Birth			Sample	Sample	Sample	Sample	
2	Number	Weights			of 3	of 3	of 30	of 30	
3	1	7.592116			6.830568	6.706177	6.65156	7.331414	
4	2	6.160641			7.599144	6.751228	7.905913	6.710266	
5	3	6.925447			7.108211	6.96647	6.809353	6.925447	
6	4	6.672467		Mean	7.179308	6.807958	7.077121	7.486474	
7	5	7.283647		StdDev	0.38919	0.139111	7.066425	7.087193	
8	6	6.818563					7.063442	8.192396	
31	29	6.929912					7.085614	7.704375	
32	30	6.914709					7.084522	7.293302	
33	31	6.233269				Mean	7.055496	6.962239	
34	32	6.505549				StdDev	0.345667	0.408372	
35	33	5.879552							
36	34	6.865114	% of dif mean		5.2%		1.3%	% of d f mean	
37	35	6.706177	%of diff StdDev		64.3%		18.1%	%of diff StdDev	
38	36	7.169123							

Each worksheet will be different. In the example above you can see that between the two samples of 3 each, the percent of difference in the means was 5.2% and the percent of difference in the standard deviation was 64.3%. Between the two samples of 30 each, the percent of difference in the means was only 1.3% and the percent of difference in the standard deviation was only 18.1%. There was more difference in the means and standard deviation of the smaller samples than of the larger samples. Also in the two samples of 30, the means, 7.06 and 6.96 are closer to the population mean of the birth weights (7.04). Every sample will be different, but in general the larger samples should have less variation.

If you wish to show the results of this exercise, print only the first page of your worksheet by doing the following:

1. Click the Office button and select Print.

2. In the Print dialog box, under Print range, select Page(s) radio button. In the From text box, key **1**. In the To text box, key **1**. Click OK.

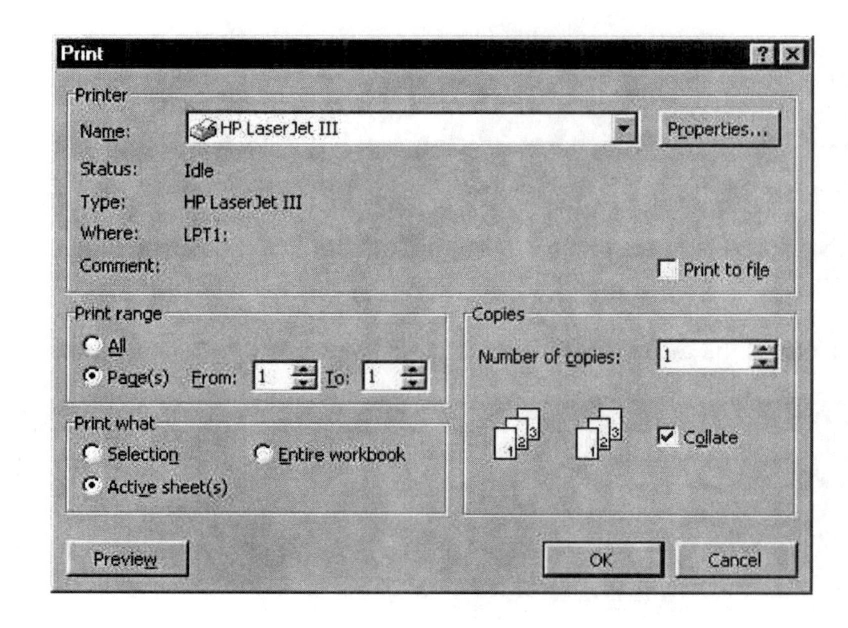

The Print icon cannot be used to specify a print range of pages from a multiple page document.

If you wish, save your file as **Ch4-examp2-b**. Close your file.

c. Plotting of Sample Means

Construct histograms using a very small sampling distribution of sample means and a larger sampling distribution of sample means.

1. On a new worksheet, enter the data as shown below.

	A	B	C	D	E	F	G
1	Sample	Sample Values			Sample		
2	Number	1	2	3	Mean	Class	Freq
3	1						
4	2						
5	3						
6	4						
7	5						
8	6						
9	7						
10	8						
11	9						
12	10						
13	11						
14	12						
15							

2. Make B3 your active cell. Key =**NORMINV(RAND(),7.04,.42)** Press <Enter>.

3. Make B3 your active cell. Place your mouse arrow on the lower right handle of B3. Make sure you have a thick, black plus sign. Click your left mouse button twice, rapidly.

4. B3:B14 should be highlighted. Place your mouse arrow on the lower right handle of B14. Drag B14 to D14.

This generates random numbers.

3. In E3, key =**AVERAGE(B3:D3).** Press <Enter>.

4. Using the method in step 3, copy E3 to E4:E14.

This is the mean of each sample.

7. In F3 and F4, key **5.9** and **6.15**, respectively.

8. Highlight **F3:F4**. Place your mouse pointer on the lower right handle of F4. Drag to F14.

This increases the weights in increments of .25 between 5.9 and 8.65, giving the upper class boundaries for the frequency distribution.

9. Highlight G3:G14. With the range still highlighted, key =**FREQUENCY(E3:E14,F3:F14)** <u>DO NOT PRESS THE <ENTER> KEY YET!</u>

10. After you have finished keying, hold down the <Shift> key and the <Ctrl> key together and at the same time press the <Enter> key. The formula in the formula bar at the top of the worksheet should be inside curly brackets, { }.

This is called an *array*. An array links the data together and prevents the formula from being accidentally over-written.

Your worksheet should be similar to the one below. The numbers will be different but the format should be the same.

G3 ▾ fx {=FREQUENCY(E3:E14,F3:F14)}

	A	B	C	D	E	F	G	H
1	Sample	Sample Values			Sample			
2	Number	1	2	3	Mean	Class	Freq	
3	1	6.560988	7.482278	7.344474	7.129247	5.9	0	
4	2	7.252956	6.889617	6.764527	6.969033	6.15	0	
5	3	7.523549	7.063557	6.877037	7.154714	6.4	0	
6	4	7.378979	7.265766	7.16661	7.270452	6.65	1	
7	5	6.613969	7.023182	7.500672	7.045941	6.9	2	
8	6	6.380031	6.845983	6.625309	6.617107	7.15	5	
9	7	7.136964	6.787397	7.732761	7.21904	7.4	3	
10	8	7.363295	7.019149	6.795638	7.059361	7.65	1	
11	9	7.027907	7.618707	6.376304	7.007639	7.9	0	
12	10	6.788718	6.751032	6.965526	6.835092	8.15	0	
13	11	7.141421	7.277546	5.895645	6.771538	8.4	0	
14	12	7.424651	7.194596	7.766088	7.461778	8.65	0	
15								

Press the F9 function key on your keyboard. The numbers in the frequency column should change. You will now plot the frequency distribution on a histogram.

1. From the Insert tab, select Column chart. Choose the upper left column chart. Click Next.
2. Under the Design tab, click Select Data. The Select Data Source dialog box will appear. Select Add under Legend Entry (Series).
3. In the Edit Series dialog box:

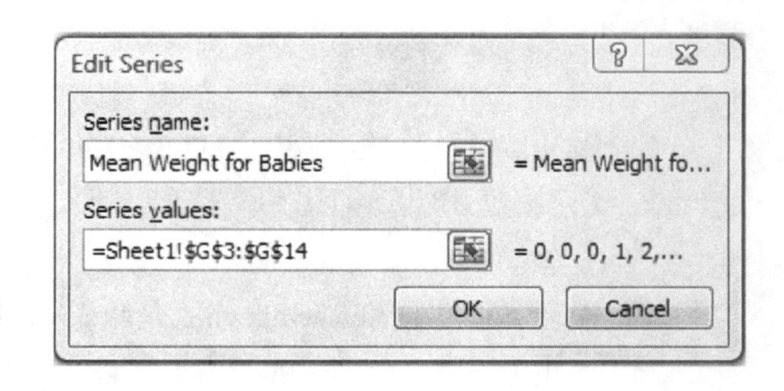

   a. Key **Mean Weights for Babies,** under Series name.
   b. Click the icon to the right of the Series values text box. Put the cursor on cell G3, click hold and drag to cell G14.
   c. Click OK.
4. Click Add under Horizontal (Category) Axis Labels.
5. In the Axis Labels dialog box, click the icon to the right of the text box. Put the cursor on cell F3, click, hold and drag to cell F14. Press Enter.
6. Click OK.

7. Under the Layout tab select Axis Titles. Select Primary Horizontal Axis Title and Title Below Axis. Key **Interval Weights** and press Enter.
8. Select the Primary Vertical Axis Title and Rotated Axis. Key **Frequency** and press Enter.
9. Select the legend on the chart and press the delete key.
10. Make sure the frame shows around the chart. With your **right** mouse arrow, click on one of the columns. Select Format Data Series.
11. The Series Options tab should be selected. In the Gap width text box, drag on the arrow until the Gap width reads 0. Click Close.

A condensed chart is displayed below.

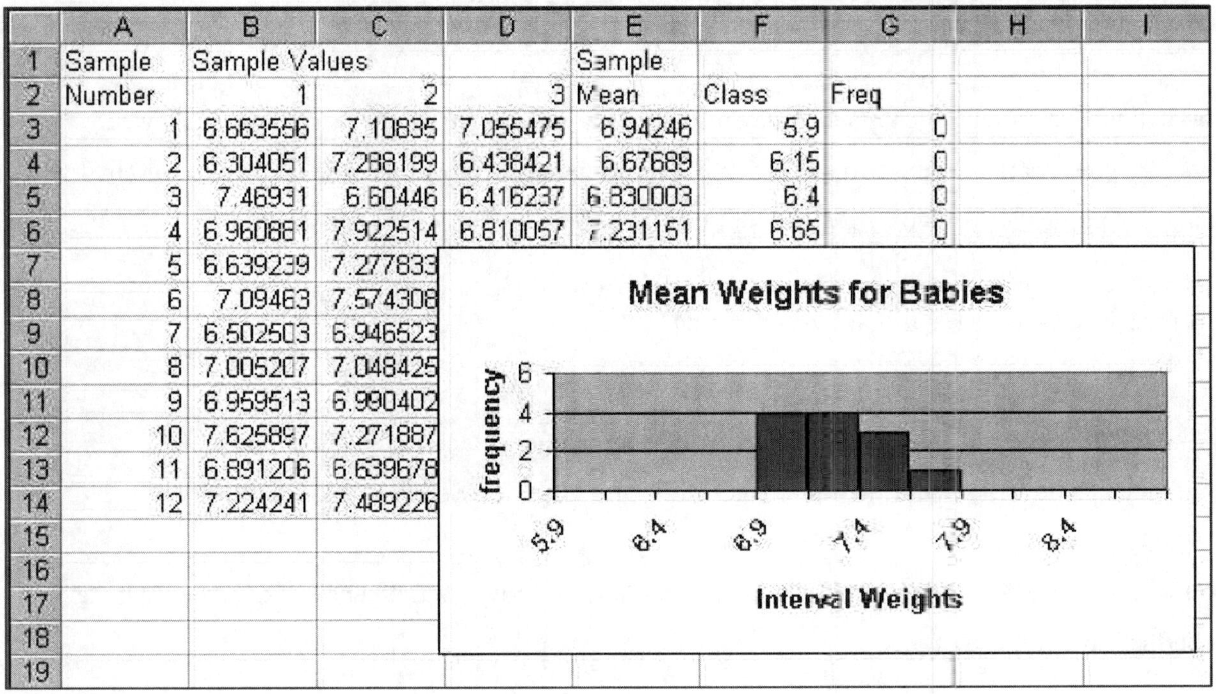

	A	B	C	D	E	F	G	H	I
1	Sample	Sample Values			Sample				
2	Number	1	2	3	Mean	Class	Freq		
3	1	6.663556	7.10835	7.055475	6.94246	5.9	0		
4	2	6.304051	7.288199	6.438421	6.67689	6.15	0		
5	3	7.46931	6.60446	6.416237	6.830003	6.4	0		
6	4	6.960881	7.922514	6.810057	7.231151	6.65	0		
7	5	6.639239	7.277833						
8	6	7.09463	7.574308						
9	7	6.502503	6.946523						
10	8	7.005207	7.048425						
11	9	6.959513	6.990402						
12	10	7.625897	7.271887						
13	11	6.891206	6.639678						
14	12	7.224241	7.489226						
15									
16									
17									
18									
19									

Mean Weights for Babies

frequency

6
4
2
0

5.9
6.4
6.9
7.4
7.9
8.4

Interval Weights

12. With the frame still around the chart, click and hold the left mouse button inside the chart. A 4-way arrow will show in the chart. As you move the chart it will show as an open box. With your mouse button still depressed, drag your mouse and move your chart so the left edge of the chart is in column H and the top edge of the chart is in row 2.
13. Click on the bottom handle of the chart. Drag the bottom line to row 16.

Press the F9 function key on your keyboard. The histogram will change according to the random weights selected by the computer.

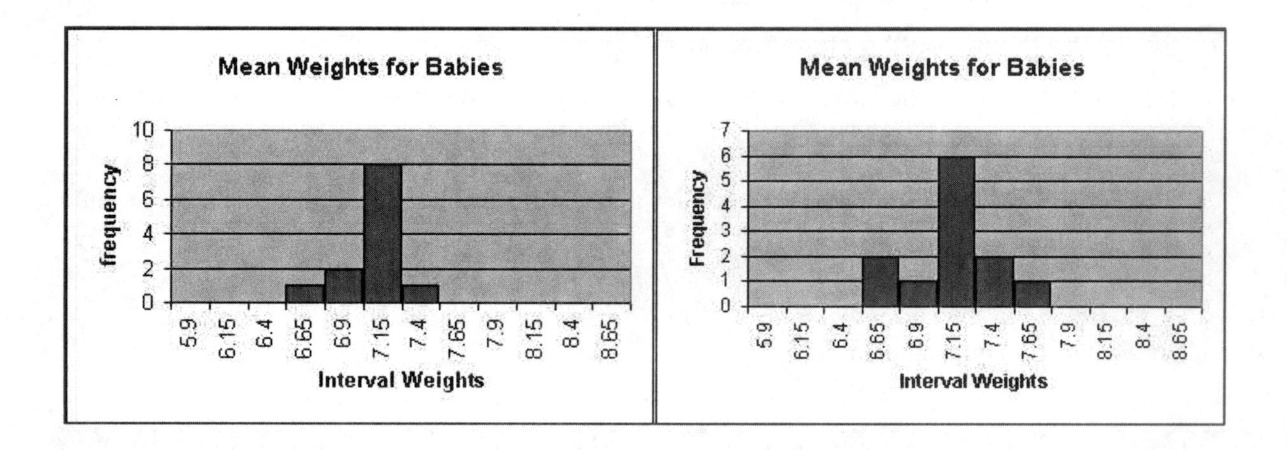

You will have a variety of charts that may resemble a normal curve, or they may appear skewed. Compare your chart with the two sample charts shown above. As you press the F9 key your chart will change. At times it may look very different from the above example, or very similar.

If you wish, save your file as **Ch4-examp2-c1**

To unhide the columns, click on the Select worksheet button located to the left of column A . Under the Home tab select Format. Select Hide and Unhide. Select Unhide Columns.

You will now insert additional columns so you can observe how the frequency of the means changes when there is a larger sample.

1. Click on the C of column C and drag to column AC. This will highlight columns C to AC. From the Home tab, select Insert. Select Insert Sheet Columns.

This inserts extra columns between the existing columns so you will not have to re-enter the formulas for the mean and frequency, nor re-do the chart.

2. Make B2 your active cell. From the Home tab, select Fill. Select Series. From the Series dialog box, select Series in Rows. Select Type Linear. In the Step Value text box, key **1**. In the Stop Value text box, key **28**. Click OK.

3. In cell AD2:AE2, key **29** and **30**, respectively.

4. Highlight B3:B14. Drag the lower right handle of B14 to AC14.

Click on the bottom scroll bar until your histogram is visible. Push the F9 function key several times.

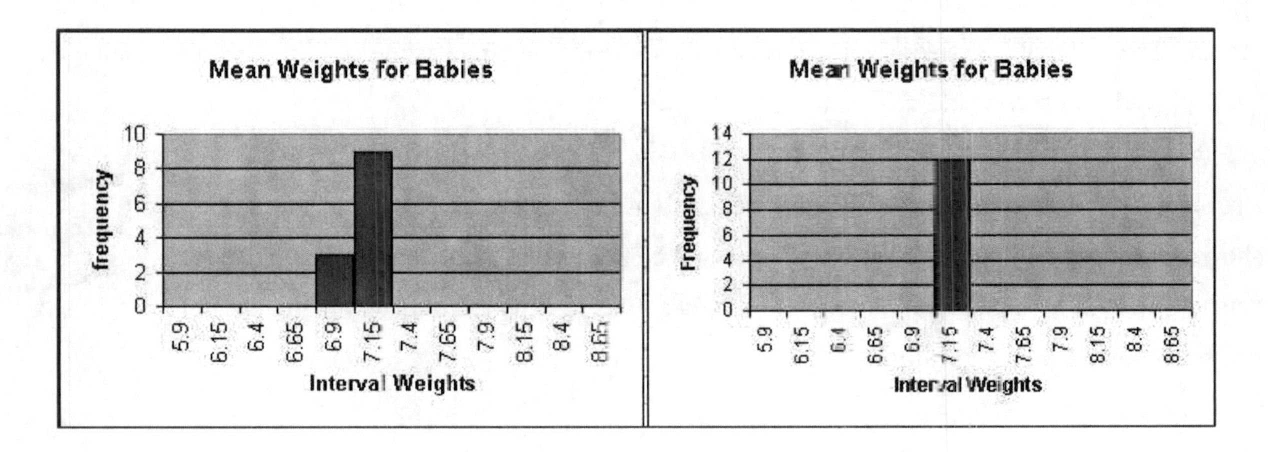

Notice how the histogram changes.

Compare your chart with the two sample charts shown above. Even though your chart changes, the bars are pretty much centered around the middle because taking the mean of several means reduces the amount of dispersion.

> If you wish to show the results of this exercise, you can print only the first page of your worksheet by doing the following. Highlight A1:AD1. From the Home tab, select Format. Select Hide & Unhide. Select Hide Columns. This hides part of your worksheet so you can print more easily. Click the Office button, select Print. Select Print again. From the Print dialog box, select Page(s). In the From text box, key **1**. In the To text box, key **1**. Click OK.

If you wish, save your file as **Ch4-examp2-c2**. Close your file.

Counting Rules

The formulas for the counting techniques of finding combinations and permutations are built into Excel®. Excel®, therefore, is useful to find how many subsets can be obtained from a set. In selecting the elements in the subsets, the distinction between combinations and permutations depends on whether the order of the selection makes a difference.

When you use the paste function to find permutations and combinations, the dialog box fills most of the screen. The cell that is active when you access the different functions is the cell in which the results will be displayed. Be sure to identify your work.

Problem 4-1: A television news director wishes to use three news stories on an evening show. One story will be the "lead story," one will be the second story, and the last will be a "closing story." If the director has a total of eight stories to choose from, how many possible ways can the program be set up?

On a new worksheet, key the following as shown.

	A	B	C	D
1	Permutation			
2	n=8			
3	r=3			
4		different ways		
5				
6	Combination			
7	n=12			
8	r=5			
9		different combinations		
10				

Make A4 your active cell. From the Formula bar, click on the Insert Function icon.

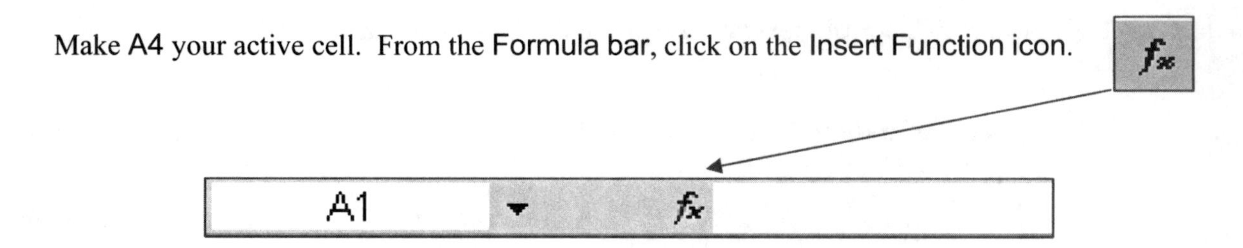

The Insert Function dialog box is displayed.

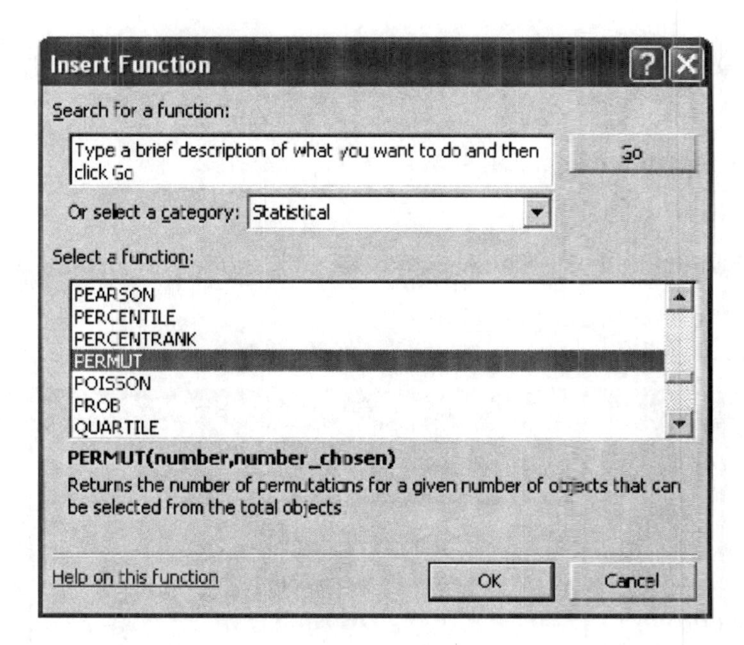

1. Click your mouse arrow on the down arrow of the Or select a category: scroll bar. Select Statistical.

2. Click your mouse arrow on the down arrow of the Select a function: scroll bar. Select PERMUT.

3. Click OK.

The second dialog box contains several text boxes to fill.

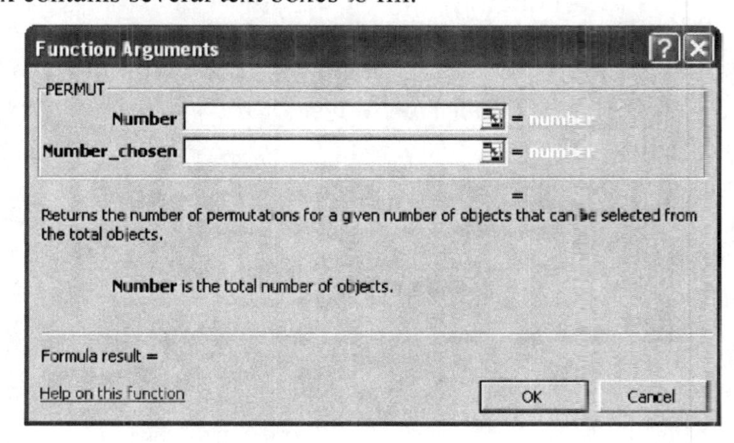

The ***Number*** text box is the number of objects, commonly referred to as *n*.

The ***Number_chosen*** text box is the number of objects in each permutation, commonly referred to as *r*.

4. Your cursor should be on the Number text box. Key **8**. Touch the tab key.

5. In the Number_chosen text box, key **3**

As soon as you enter the number in the Number_chosen text box, the permutation shows after *Formula result* = in the lower left corner of the dialog box.

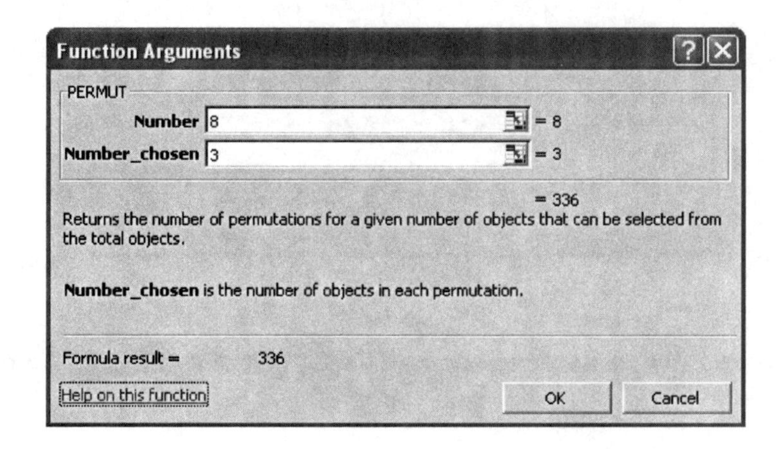

6. Click OK.

The permutation of 336 is displayed in cell A4.

A4 ▾ fx =PERMUT(8,3)

	A	B	C	D
1	**Permutation**			
2	n=8			
3	r=3			
4	**336**	different ways		
5				
6	Combination			
7	n=12			
8	r=5			
9		different ways		
10				

Bold the contents of A1 and A4.

Problem 4-2: A bicycle shop owner has 12 mountain bicycles in the showroom. The owner wishes to select 5 of them to display at a bicycle show. How many different ways can a group of 5 be selected?

1. On the same worksheet make A9 your active cell. From the Formula bar, click on the Insert Function icon.

2. From the Or select a category: list box, select Math & Trig.

3. From the Select a function: scroll bar, select COMBIN. Click OK.

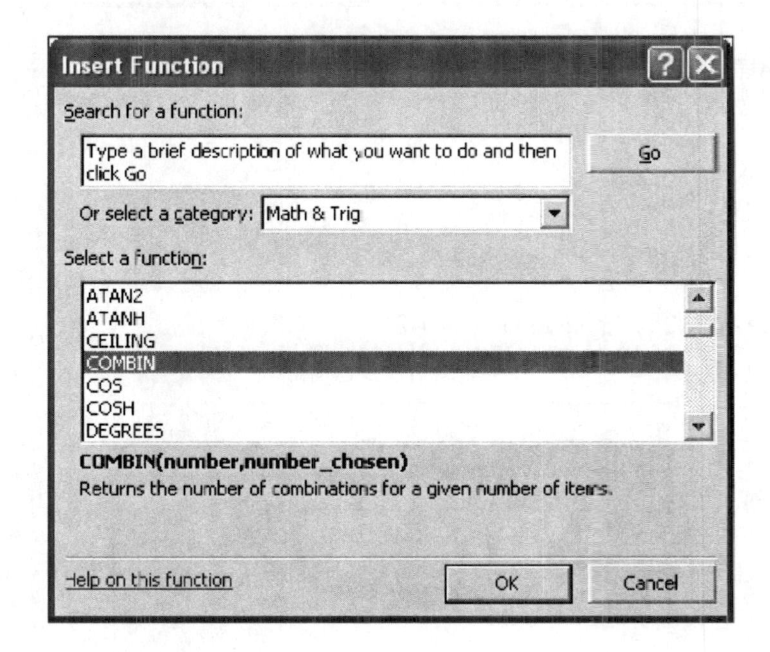

The second dialog box contains two text boxes to fill.

The ***number*** text box is the number of items commonly referred to as n.

The ***number_chosen*** text box is the number of objects in each combination, commonly referred to as r.

4. Your cursor should be on the number text box. Key **12**. Press the tab key.

5. In the number_chosen text box, key **5**.

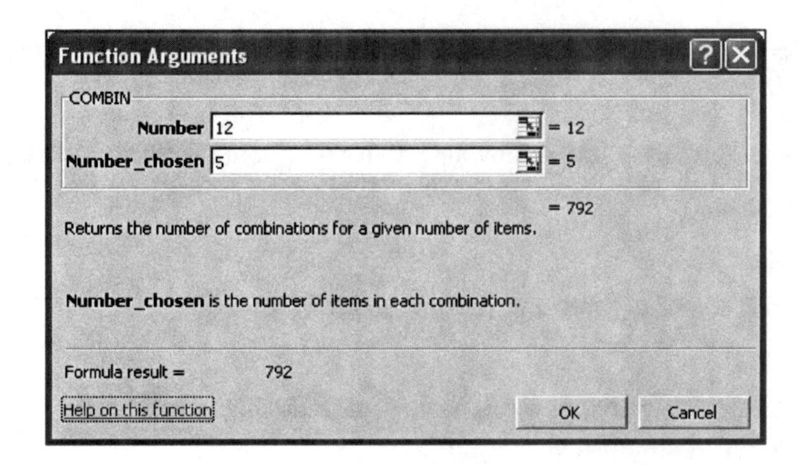

As soon as your last data is entered, the combination shows after *Formula result* = in the lower left corner of the dialog box.

6. Click OK.

The combination value of 792 is displayed in cell A9.

A9		f_x =COMBIN(12,5)	
	A	B	C	D

	A	B	C	D
1	**Permutation**			
2	n=8			
3	r=3			
4	**336**	different ways		
5				
6	**Combination**			
7	n=12			
8	r=5			
9	**792**	different ways		
10				

Bold the contents of A6 and A9.

If you wish, save your file as **Ch4-probs**

You can put several different problems on one worksheet. Just be sure the cell in which you want the result is the active cell when you access the Insert Function.

Practice Exercises taken from textbook.

4-1. In a board of directors composed of 8 people, how many ways can a chief executive officer, 1 director, and 1 treasurer be selected? (Textbook Exercise 4-5, Problem 20)

4-2. The County Assessment Bureau decides to reassess homes in8 different areas. How many different ways can this be accomplished? (Textbook Exercise 4-5, Problem 14)

4-3. An inspector must select 3 tests to perform in a certain order on a manufactured part. He has a choice of 7 tests. How many ways can he perform 3 different tests? (Textbook Exercise 4-5, Problem 16)

4-4. How many different ways can 4 tickets be selected from 50 tickets if each ticket wins a different prize? (Textbook Exercise 4-5, Problem 23)

4-5. How many ways are there to select 3 bracelets from a box of 10 bracelets disregarding the order of selection? (Textbook Exercise 4-5, Problem 29)

4-6. How many ways can a committee of 4 people be selected from a group of 10 people? (Textbook Exercise 4-5, Problem 31)

4-7. How many different tests can be made from a test bank of 20 questions if the test consists of 5 questions? (Textbook Exercise 4-5, Problem 33)

4-8. A candy store allows customers to select 3 different candies to be packaged and mailed. If there are 13 varieties available, how many possible selections can be made? (Textbook Chapter 4, Review Exercise #36)

CHAPTER 5
DISCRETE PROBABILITY DISTRIBUTIONS

Chapter 2 showed you how you can use Excel® to graphically portray data in pie charts, bar charts, histograms, line charts, etc. Chapters 3 & 4 emphasized descriptive statistics with measurements of central tendency and dispersion. Those chapters were concerned with descriptive statistics, describing something that had already occurred. Chapter 4 also introduced you to the second facet of statistics, that is, computing the chance or probability that something *will* occur. In this chapter you will continue to use Excel® Insert Function tool to find three different **discrete distributions**:

- Use Excel® to create discrete probability distributions
- Use Excel® to find binomial probabilities
- Use Excel® to find hypergeometric probabilities
- Use Excel® to find Poisson distributions

Here are a few examples of discrete probability distributions and how they are used.

Binomial Distribution

When you use Excel® to find various statistical functions, the dialog box fills most of the screen. **The cell that is active when you access the different functions is the cell in which the result will be displayed.**

1. On a new worksheet, key the following as shown:

	A	B	C
1	Binomial Distributions		
2	a. x=5		
3	b. x<=3		
4			
5	c. x>=3		
6			

2. Make B2 your active cell.

To use the binomial distribution function of Excel®, do the following:

3. From the Formula bar, click on the Insert Function icon. 

The Insert Function dialog box is displayed.

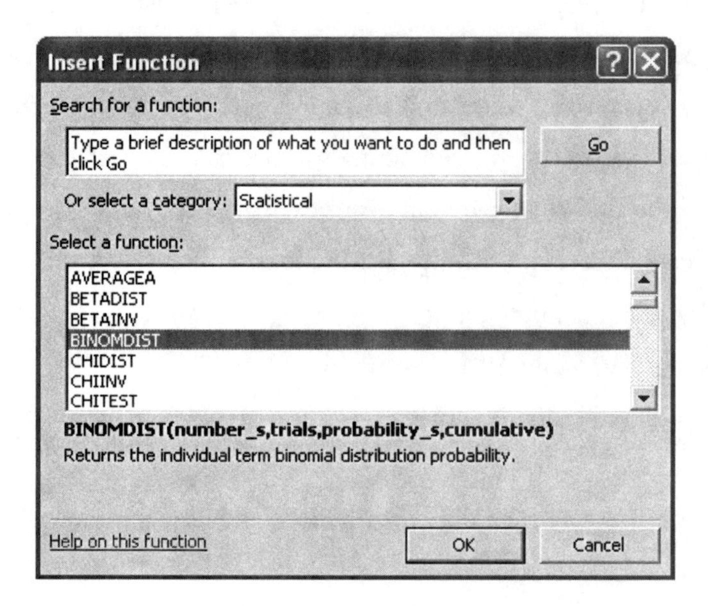

4. In the Or select a category list box, select Statistical.

5. From the Select a function scroll bar, select BINOMDIST.

6. Click on OK.

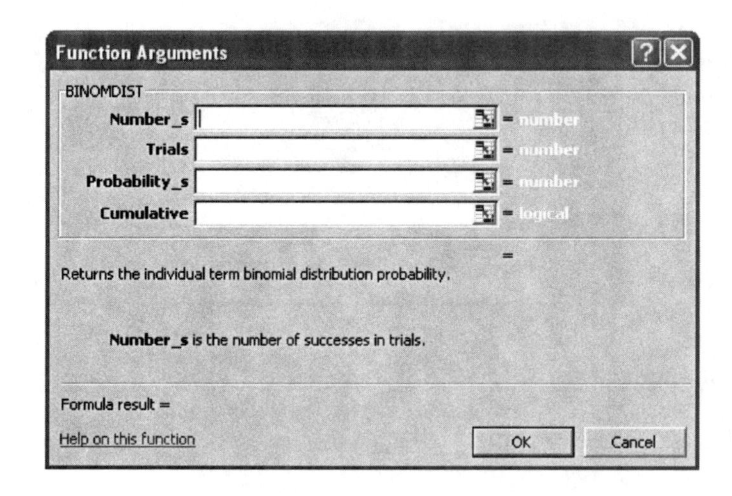

The BINOMDIST dialog box contains several text boxes to fill, as described below:

- The ***Number_s*** text box is for the number of successes in trials, commonly referred to as *x*.
- The ***Trials*** text box is for the number of independent trials, commonly referred to as *n*.
- The ***Probability_s*** text box is for the probability of success on each trial, commonly referred to as *p*.
- The ***Cumulative*** text box is used to indicate whether *x*, the number of observed successes, is cumulative or not cumulative. You would key 1 for true if you want the cumulative probability that includes the numbers up to and including the value of *x*. You would key 0 for false (or not cumulative) if *x* is the number of observed successes only.

You will use this dialog box to solve part (a) of the following problem:

Problem 5-1. It is reported that 5 % of Americans are afraid of being alone in a house at night. If a random sample of 20 Americans is selected, find these probabilities:

a. There are exactly 5 people in the sample who are afraid of being alone at night.
b. There are at most 3 people in the sample who are afraid of being alone at night.
c. There are at least 3 people in the sample who are afraid of being alone at night.

Example a. In the binomial distribution, x=5, n = 20 and p = .05.

1. Your cursor should be on the Numbers_s text box. Key **5**. Press the tab key.
2. In the Trials text box, key **20**. Press the tab key.
3. In the Probability_s text box, key **.05** . Press the tab key.
4. In the Cumulative text box, since you want exactly 5 observed success, key **0** for false (not cumulative).

As soon as you enter the number in the Cumulative text box, the probability shows after the "Formula result =" in the lower left corner of the dialog box.

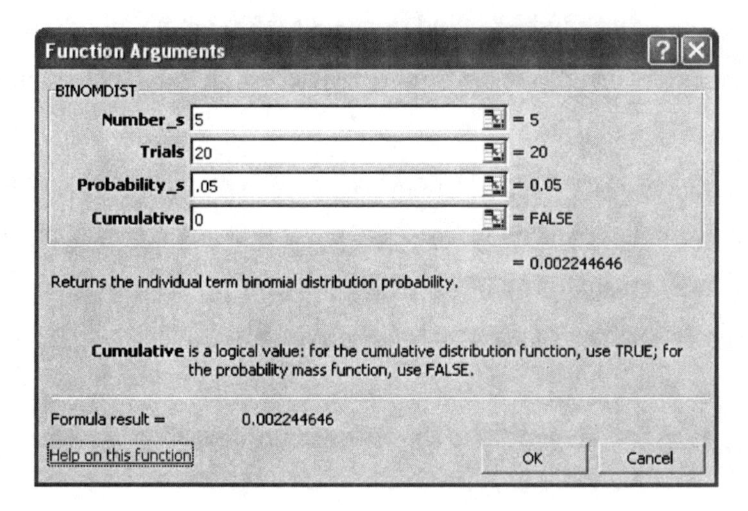

5. Click OK.

6. Make B2 your active cell. From the Home tab select the Decrease Decimal icon. Click until the decimal is rounded to three decimal places.

The probability value of .002 is displayed in cell B2.

B2 f_x =BINOMDIST(5,20,0.05,0)

	A	B	C	D	E
1	Binomial Distributions				
2	a. x=5	0.002			
3	b. x<=3				
4					
5	c. x>=3				
6					

To complete example b of the problem, make B3 your active cell.

Example b. In the binomial distribution, x = 3, n = 20 and p =. 05

1. Follow the instructions for inserting the BINOMDIST function.

2. Your cursor should be in the number_s text box. Key **3**. Press the tab key.

3. In the trials text box, key **20**. Press the tab key.
4. In the probability_s text box, key **.05**. Press the tab key.
5. Your cursor should be in the cumulative text box, since you want to select cumulative (0 through 3 inclusive), key **1** for true. Click OK.
6. Decrease the decimal to three decimal places.

The probability value of .984 is now displayed in cell B3.

	A	B
1	Binomial Distributions	
2	a. X=5	0.002
3	b. X<=3	0.984
4		

To complete example c of the problem, make B4 your active cell. Since the probability of at least three people is **not** two or less people, you will find the probability for x being 2 cumulative, and then find the complement (subtracted from 100 %).

Example c. In the binomial distribution, x = 2, n = 20 and p = .05

1. Follow instructions for using the BINOMDIST function.
2. Your cursor should be in the number_s text box. Key **2**. Press the tab key.
3. In the trials text box, key **20**. Press the tab key.
4. In the probability_s text box, key **.05**. Press the tab key.
5. Your cursor should be in the cumulative text box, since you want to select cumulative (0 through 2 inclusive), key **1** for true. Click OK.
6. Decrease the decimal to three decimal places.

The probability value of .925 is now displayed in cell B4.

7. Make cell B5 your active cell. Key in the formula =**1-B4**. Press the <Enter> key. Decrease the decimal to three decimal places. The value .075, which is the compliment of $x \leq 2$, is displayed. This is the answer to Example c of the problem.

8. Bold cells A2:B3 and cells A5:B5 to make the problems and answers easily identifiable.

B5 ▾ fx =1-B4

	A	B	C	D
1	Binomial Distributions			
2	**a. x=5**	**0.002**		
3	**b. x<=3**	**0.984**		
4		0.925		
5	**c. x>=3**	**0.075**		
6				

You can use the BINOMDIST function and the complement of the probability (subtracted from 1) to solve many binomial distributions.

Poisson Distribution

Problem 5-2. If approximately 2% of the people in a room of 200 people are left-handed, find the probability that exactly five people there are left-handed.

On the same worksheet, in cell A7 key **Poisson Distribution**. In Cell A8 key **x=5.** Make cell B8 your active cell.

Before you start this problem you need to find the mean number of left-handed people out of 200. There is an average of 2% of 200, or a mean of 4. The Greek letter λ (lambda) is used to represent the mean number of occurrences per unit (area, time, volume, etc.).

To use the Poisson distribution function of Excel®, do the following:

1. From the Formula bar, click on Insert Function.

2. From the Or select a category list box, select Statistical.

3. Click your mouse arrow on the down arrow of the Select a function scroll bar. Select POISSON.

4. Click OK.

Step 2 of the POISSON dialog box contains several text boxes to fill.

- The ***X*** text box is for the number of events (successes).
- The ***Mean*** text box is for the expected numeric value. This is the arithmetic mean number of occurrences (successes) in a particular interval of time, commonly referred to as λ (lambda).
- The ***Cumulative*** text box is used to indicate whether *X*, the number of occurrences (successes) is cumulative or not cumulative. You would key 1 for true (cumulative) or 0 for false (not cumulative).

You will use this dialog box to continue this problem.

5. Your cursor should be in the X text box Key **5**. Press the tab key.

6. In the **Mean** text box, key **4**. Press the **tab** key.

7. In the **Cumulative** text box, key **0**.

8. Click **OK**.

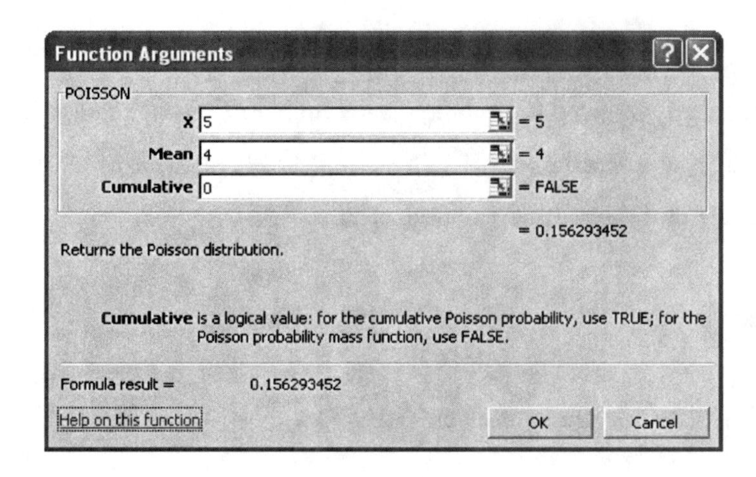

9. Decrease the decimal to three decimal places.

The probability of .156 is now displayed in cell B8.

B8 fx =POISSON(5,4,0)

	A	B	C	D	E
1	Binomial Distributions				
2	**a. x=5**	**0.002**			
3	**b. x<=3**	**0.984**			
4		0.925			
5	**c. x>=3**	**0.075**			
6					
7	**Poisson Distribution**				
8	x=5	**0.156**			
9					

10. Bold cells A8:B8.

As with the binomial distribution, you can use 0 or 1 in the cumulative text box, and use the compliment to find the probability for various values of x.

Hypergeometric Distribution

Problem 5-3. A recent study found that 4 out of nine houses were underinsured. If five houses are selected from the nine houses, find the probability that exactly two are underinsured.

On the same worksheet, in cell A10, key **Hypergeometric Distribution**. In A11 key **x=2**. Make B11 your active cell. To use the hypergeometric distribution function of Excel®, do the following.

1. From the Formula bar, click on Insert Function.

2. From the Or select a category list box, select Statistical.

3. Click your mouse arrow on the down arrow of the Select a function scroll bar. Select HYPGEOMDIST.

4. Click OK.

The HYPGEOMDIST dialog box contains several text boxes to fill.

- The ***Sample_s*** text box is for the number of successes in the sample, commonly referred to as X.
- The ***Number_sample*** text box is for the size of the sample, commonly referred to as *n*.
- The ***Population_s*** text box is for the number of successes in the population, referred to as *S*.
- The ***Number_pop*** text box is for the population size, referred to as *N*.

You will use this dialog box to continue this problem.

5. Your cursor should be on the Sample_s text box. Key **2**. Press the tab key.

6. In the Number_sample text box, key **5**. Press the tab key.

7. In the Population_s text box, key **4**. Press the tab key.

8. In the Number_pop text box, key **9**.

9. Click OK.

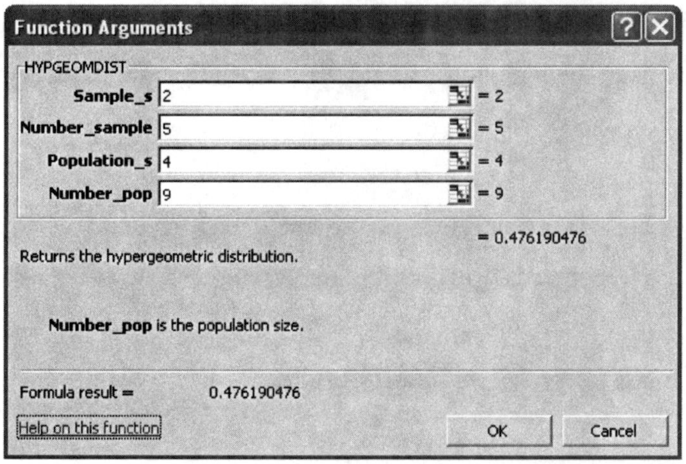

Function Arguments

HYPGEOMDIST

Sample_s 2 = 2

Number_sample 5 = 5

Population_s 4 = 4

Number_pop 9 = 9

= 0.476190476

Returns the hypergeometric distribution.

Number_pop is the population size.

Formula result = 0.476190476

Help on this function

OK Cancel

10. Decrease the decimal to three decimal places.

The probability value of .476 is now displayed in cell B11.

11. Bold cells A11:B11

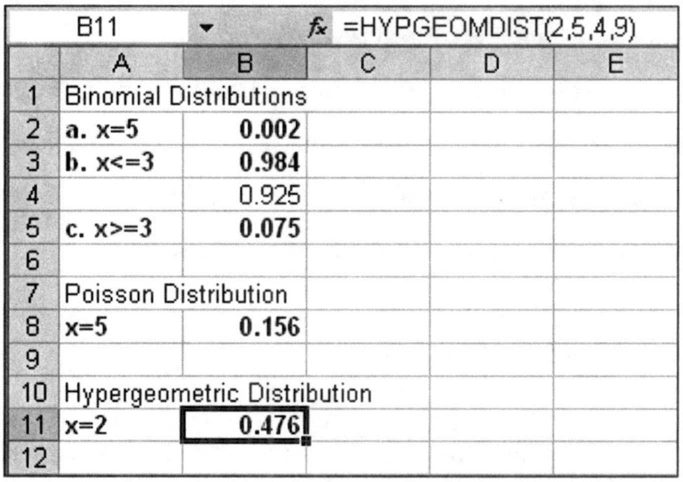

B11 fx =HYPGEOMDIST(2,5,4,9)

	A	B	C	D	E
1	Binomial Distributions				
2	**a. x=5**	**0.002**			
3	**b. x<=3**	**0.984**			
4		0.925			
5	**c. x>=3**	**0.075**			
6					
7	Poisson Distribution				
8	x=5	**0.156**			
9					
10	Hypergeometric Distribution				
11	x=2	**0.476**			
12					

Save your file as **Chap 5 probs.** Print the probabilities if you wish. Close your worksheet.

Practice Exercises taken from textbook. Use discrete distributions to solve the following problems. Be sure to identify your outcome on the worksheet. Also remember that the cell that is active when you work the exercises is the cell in which the results will be displayed.

5-1. A burglar alarm system has 6 fail-safe components. The probability of each failing is .05. Find these probabilities. (Textbook Exercise 5-4, Problem 4)

a. Exactly three will fail.
b. Fewer than two will fail.
c. None will fail.

5-2. R. H. Bruskin Associates Market Research found the 40% of Americans do not think that having a college education is important to succeed in the business world. If a random sample of five Americans is selected, find these probabilities.. (Textbook Exercise 5-4, Problem 1[illegible])

a. Exactly two people will agree with that statement.
b. At most three people will agree with that statement.
c. At least two people will agree with that statement
d Fewer than three people will agree with that statement

5-3. If 30% of all commuters ride the train to work, find the probability that if 10 workers are selected, 5 will ride the train. (Textbook Chapter 5, Review Exercise # 14)

5-4. If 10% of the people who are given a certain drug experience dizziness, find these probabilities for a sample of 15 people who take the drug. (Textbook Chapter 5, Review Exercise # 16)

a. At least two people will become dizzy.
b. Exactly three people will become dizzy.
c. At most four people will become dizzy.

5-5. In a 400-page manuscript, there are 200 randomly distributed misprints. If a page is selected, find the probability that it has one misprint. (Textbook Exercise 5-5, Problem 10)

5-6. A mail-order company receives an average of five orders per 500 solicitations. If it sends out 100 advertisements, find the probability of receiving at least two orders. (Textbook Exercise 5-5, Problem 12)

5-7. If 4% of the population carries a certain genetic trait, find the probability that in a sample of 100 people, there are exactly 8 people who have the trait. Assume the distribution is approximately Poisson. (Textbook Chapter 5, Review Exercise # 26)

5-8. The number of boating accidents on Lake Emilie follows a Poisson distribution. The probability of an accident is .003. If there are 1000 boats on the lake during a summer month, find the probability that there will be 6 accidents. (Textbook Chapter 5, Review Exercise # 28)

5-9. Shirts are packed at random in two sizes, regular and extra large. Four shirts are selected from a box of 24 and checked for size. If there are 15 regular shirts in the box, find the probability that all 4 will be regular size. (Textbook Exercise 5-5, Problem 19)

5-10. If five cards are drawn from a deck, find the probability that two will be hearts. (Textbook Chapter 5, Review Exercise # 29)

5-11. Of the 50 automobiles in a used-car lot, 10 are white. If five automobiles are selected to be sold at an auction, find the probability that exactly two will be white. (Textbook Chapter 5, Review Exercise # 30)

CHAPTER 6
THE NORMAL DISTRIBUTION

Chapter 5 dealt with discrete probability distributions. This chapter will continue the study of probability distributions by examining a very important continuous probability distribution- the normal distribution, and the concept of the Central Limit theorem. In this chapter you will:

- Use Excel® to create a normal probability distribution.
- Use Excel® and the standard normal distribution to determine the probability that an observation will be above or below a value.
- Use Excel® and the standard normal distribution to determine the probability that an observation will lie between two points.
- Use Excel® and the standard normal distribution to find the value of an observation when the percent above or below the observation is given.
- Use Excel® to generate a frequency distribution and to illustrate the central limit theorem.

Areas Under the Normal Curve

To use Excel® to compute the area under a normal curve, you will be creating a template. By pasting sections of the template onto a new worksheet, you can solve a variety of problems. The template can be used with a value (X) and a given mean and standard deviation, or can be used for z-values assuming the mean is 0 and the standard deviation is 1. Although the textbook lists 7 types of problems, by using Excel®'s normal cumulative distribution, you can categorize the problems into 5 types. Type 1 is an area under the curve that is less than a particular value (X). Type 2 is an area under the curve that is between a particular value (X) and the mean, where X is less than the mean. Type 3 is an area under the curve that is between a particular value (X) and the mean, where X is greater than the mean. Type 4 is an area under the curve that is greater than a particular value (X). Type 5 is an area under the curve that is between two particular values (X's).

Please Note: When you use your text to compute the area under a curve, the given area is to the left or right of the mean. When using Excel® to compute the normal cumulative distribution, the area is cumulative from the left side of the curve. Therefore if you want to compute an area between *x* and the mean, you will compute the difference between the given area and .5 (half the area under the curve). In this exercise we will create a template that can be used for a variety of purposes.

Example 6-1. Creating a worksheet (template) to use for finding normal probability distributions.

1. On a new worksheet enter the cell contents as shown below.

	A	B	C	D	E	F	G	H
1	Normal Probability Distribution							
2								
3	Type 1				Type 4			
4								
5	Probability proportion				Probability proportion			
6	Area less than x				Area greater than x			
7								
8	Probability				CumDist=			
9	CumDist=				Probability			
10								
11								
12	Type 2				Type 5			
13								
14	Probability Proportion				Probability proportion, area between 2 x's			
15	Area between x & mean					1st x		2nd x
16	x<mean							
17					CumDist=		CumDist=	
18	CumDist=				Probability			
19	Probability							
20								
21								
22	Type 3							
23								
24	Probability proportion							
25	Area between x & mean							
26	x>mean							
27								
28	CumDist=							
29	Probability							
30								

Cells A3:B9 will be used for finding the area less than x. We will refer to them as **Type 1**. Since Excel® computes this area automatically, no other computations are needed.

Cells A12:B19 will be used when the area is between x and the mean, and x is less than the mean. We will refer to them as **Type 2**. Since Excel® computes only the area less than x, that value must be subtracted from .5 (half the area under the curve) to find the area in between.

2. In cell B19, key **=. 5-B18**

Cells A22:B29 will be used when the area is between x and the mean, and x is greater than the mean. We will refer to them as **Type 3**. Since Excel® computes only the area less than x, .5 (half the area under the curve) must be subtracted from that value.

3. In cell B29, key =**B28-.5**

Cells E3:F9 will be used when the area is more than x. We will refer to them as **Type 4**. Since Excel® computes only the area less than x, that value must be subtracted from 1 (the total area under the curve.)

4. In cell F9, key =**1-F8**

Cells E12:H18 will be used when the area is between two values of x. We will refer to them as **Type 5**. The smaller value of x is used as the first number and the larger value of x as the second number, then the two areas are subtracted.

5. In cell F18, key =**H17-F17**

When you are finished with the formulas, the cell contents of your worksheet should look as shown on the following page. To display the formula view (instead of the output), hold down the <Ctrl> key and tap the <~> key (usually located above the tab key). Use this same key combination to toggle between views.

	A	B	C	D	E	F	G	H
1	Normal Probability Distribution							
2								
3	Type 1				Type 4			
4								
5	Probability proportion				Probability proportion			
6	Area less than x				Area greater than x			
7								
8	Probability				CumDist=			
9	CumDist=				Probability	=1-F8		
10								
11								
12	Type 2				Type 5			
13								
14	Probability Proportion				Probability proportion, area between 2 x's			
15	Area between x & mean					1st x		2nd x
16	x<mean							
17					CumDist=		CumDist=	
18	CumDist=				Probability	=H17-F17		
19	Probability	=0.5-B18						
20								
21								
22	Type 3							
23								
24	Probability proportion							
25	Area between x & mean							
26	x>mean							
27								
28	CumDist=							
29	Probability	=B28-0.5						
30								

Double click on the Active worksheet tab and Key **Template**.
Save your workbook as **normprbd**. (If needed, refer to Chapter 1 on how to save.)

When you use Excel®'s Normal Distribution function, the cell that is active is the cell in which the results will be displayed. Before you choose the function always make sure your active cell is the cell to the right of the cell containing the label CumDist=.

Problem 6-1. The mean number of hours an American worker spends on the computer is 3.1 hours per workday. Assume the standard deviation is 0.5 hour. Find the percentage of workers who spend less than 3.5 hours on the computer. Assume the variable is normally distributed.

1. If it is not already open, retrieve file **normprbd**. Since you are finding an area less than x (112) you will use Type 1, the cells A3:B9 of your normprbd worksheet.

2. Highlight cells A3:B9. From your Home Tab select the Copy  icon.

3. Open a new worksheet. Make cell A3 your active cell. From the Home Tab select the Paste icon. Press the <Enter> key. 

4. In A1, key **Prob 1** to replace Type 1.

5. Make B7 your active cell since that is where you want the results to be displayed. From the Formula bar, select the Insert Function icon. 

6. From the Or select a category list box, select Statistical. Click your mouse arrow on the down arrow of the Select a function scroll bar. Select NORMDIST. Click OK.

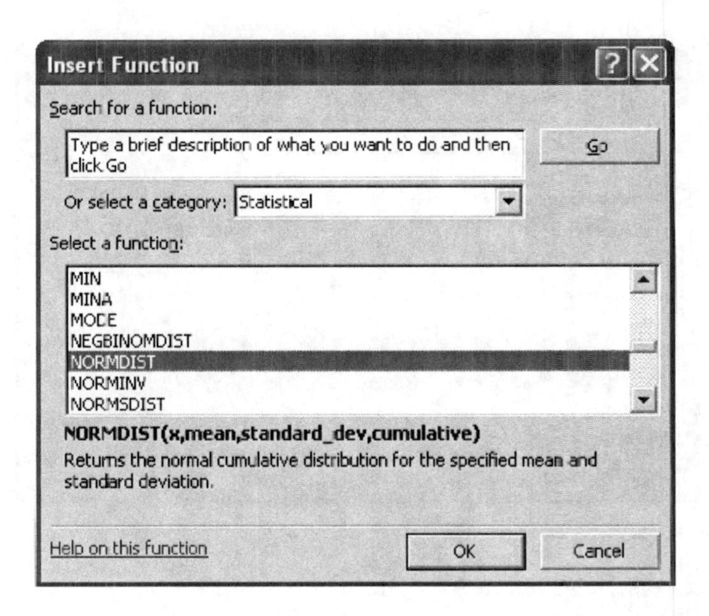

Please Note: Be careful not to confuse the Function name NORMDIST with NORMSDIST.

The dialog box for NORMDIST is displayed.

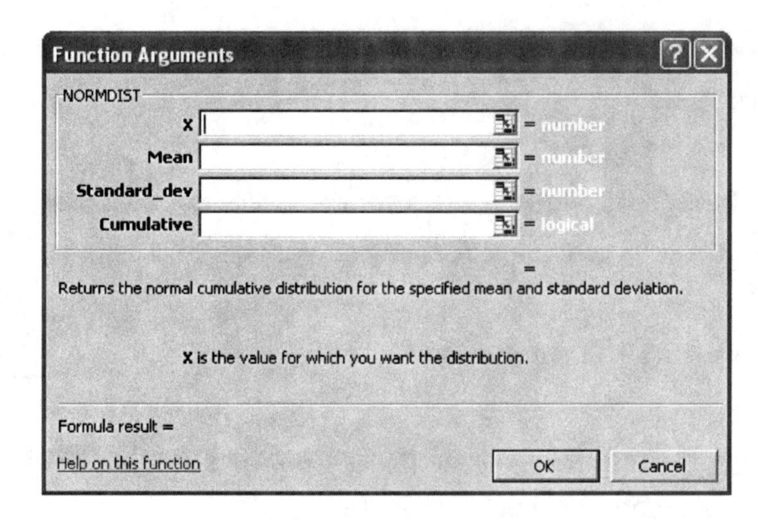

7. The cursor should be on the X text box. Key **3.5**. Press the tab key.

8. In the Mean text box, key **3.1**. Press the tab key.

9. In the Standard_dev text box, key **.5**. Press the tab key.

10. In the Cumulative text box, key **1** for true. (The cumulative text box will always contain 1) Click OK.

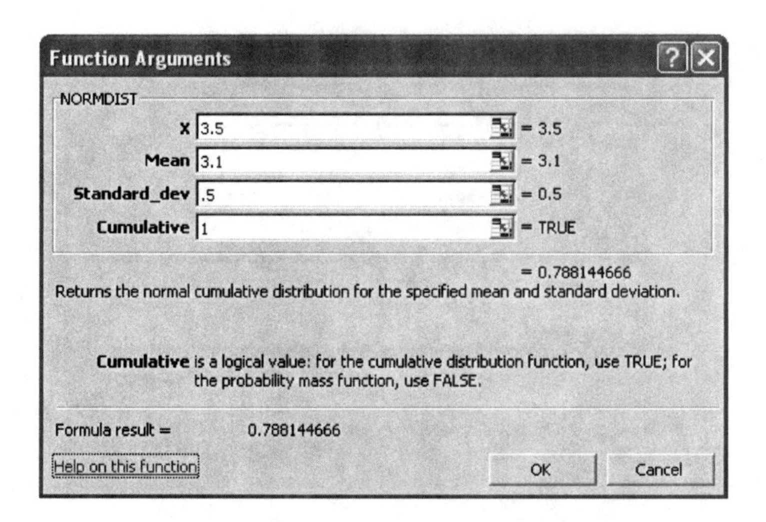

The cumulative distribution is displayed in A7. The area is .788145.

11. Under the Home tab, click on the Percent Style  icon.

12. From the Home Tab, click on the Increase Decimal icon twice to display the area as 78.81%, the percentage of scores that will fall below 3.5. 

13. Make A1 your active cell. From the Home Tab, select the Bold  icon.

14. Make B7 your active cell. From the Home Tab, select the Bold icon.

This makes the answer easier to associate with the problem.

B7 =NORMDIST(3.5,3.1,0.5,1)

	A	B	C	D	E
1	**Prob 1**				
2					
3	Probability proportion				
4	Area less than x				
5					
6	Probability				
7	CumDist=	**78.81%**			
8					

15. Double click on the active worksheet tab located at the bottom of the worksheet. Key **Problems**.

If you were to key 0 in the cumulative text box, the resulting value would be the height of the normal density function. If you were to plot the bell curve, this height would be useful for finding the value at which the curve peaks.

Problem 6-2. Each month, an American household generates an average of 28 pounds of newspaper for garbage or recycling. Assume the standard deviation is 2 pounds. If a household is selected at random, find the probability of its' generating

a. Between 27 and 31 pounds per month.
b. More than 30.2 pounds per month.

Assume the variable is approximately normally distributed.

Since part a is for an area between 2 x's (27 and 31), you will use Type 5 (cells E12:H18.)

1. On the Template worksheet Highlight cells E12:H18. From your Home Tab select the Copy icon.
2. At the bottom of the workbook make the Problems worksheet active by clicking on the tab.
3. Make cell D1 your active cell. From the Home Tab select the Paste icon. Press the <Enter> key.
4. In cell D1, key **Prob 2a**.

5. Make E6 your active cell since that is where you want the first result to be displayed. From the Formula bar, select the Insert Function icon.

6. From the Or select a category list box, select Statistical. Click your mouse arrow on the down arrow of the Select a function scroll bar. Select NORMDIST. Click OK.

The dialog box for NORMDIST is displayed.

7. The cursor should be on the X text box. Key **27**. Press the tab key.
8. In the Mean text box, key **28**. Press the tab key.

9. In the Standard_dev text box, key **2**. Press the tab key.

10. In the Cumulative text box, key **1** for true. (The cumulative text box will always contain 1) Click OK.

The cumulative distribution is displayed in E6. This is the 1st X area.

11. Make G6 your active cell since you want the results for the 2nd X to be displayed in this cell. From the Tool bar, select the Insert Function icon.

12. From the Or select a category list box, select Statistical. Click your mouse arrow on the down arrow of the Select a function scroll bar. Select NORMDIST. Click OK.

You will enter the data as before.

13. The cursor should be on the X text box. Key **31**. Press the tab key.

14. In the Mean text box, key **28**. Press the tab key.
15. In the Standard_dev text box, key **2**. Press the tab key.
16. In the Cumulative text box, key **1**. Click OK.

The cumulative distribution is displayed in G6, and the probability of the difference between the two X's is displayed in E7.

17. Make E7 your active cell. Change the value to display as a percent with two decimal places as you did in the previous problem. Bold the contents of cells D1 and E7.

	D	E	F	G
1	**Prob 2a**			
2				
3	Probability proporion, area betweer 2 x's			
4		1st x		2nd x
5				
6	CumDist=	0.308538	CumDist=	0.933193
7	Probability	**62.47%**		
8				

The probability is 62.47% that a selected household generates between 27 and 31 pounds of newspapers per month.

Part b of problem 6-2 is for an area greater than x, a Type 4 problem.

19. Make the Template worksheet the active worksheet by clicking on the Template tab.

Since part b is a Type 4, you will use cells E3:F9.

20. Highlight cells E3:F9. From your Home Tab select the Copy icon
21. Make the Problems worksheet the active worksheet by clicking on the Problems tab.
22. Make cell A10 your active cell. From the Home Tab select the Paste icon. Press the <Enter> key
23. In cell A10, key **Prob 2b**.
24. Make B15 your active cell since that is where you want this first result to be displayed. From the Formula bar, select the Insert Function icon.
25. From the Or select a category list box, select Statistical. Click your mouse arrow on the down arrow of the Select a function scroll bar. Select NORMDIST. Click OK.

You will enter data as before.

26. The cursor should be on the X text box. Key **30.2**. Press the tab key.
27. In the Mean text box, key **28**. Press the tab key.
28. In the Standard_dev text box, key **2**. Press the tab key.
29. In the Cumulative text box, key **1**. Click OK.

The cumulative distribution is displayed in B15, and the probability of the area greater than 30.2 is displayed in B16.

30. Make B16 your active cell. Change the value to display as a percent with two decimal places as you did in the previous problem.

31. Bold the contents of cells A10 and B16.

	A	B	C
10	**Prob 2b**		
11			
12	Probability proportion		
13	Area greater than x		
14			
15	CumDist=	0.864334	
16	Probability	**13.57%**	
17			

The probability is 13.57% that a selected household generates more than 30.2 pounds of newspapers per month.

You can use the normprbd template for any value of X. Just determine what type of problem it is, then copy and paste the cells you need on a new worksheet.

Please Note: Answers may vary slightly because of rounding.

Finding the Value of the Observation X when the Percent Above or Below the Observation is Given. Excel® also has a function to find the value of *x* when you know the probability, mean, and standard deviation.

Problem 6-3. In order to qualify for a police academy, candidates must score in the top 10% on a general abilities test. The test has a mean of 200 and a standard deviation of 20. Find the lowest possible score to qualify. Assume the test scores are normally distributed.

Since the must score in the top 10 percent, there will be 90 percent lower than the score, so you will use .90 as the probability. Again, the cell that is active is the cell in which the results will be displayed.

1. On the Problems worksheet, in D10, key **Prob 3**.

2. Make D11 your active cell. From the Formula bar, select the Insert Function icon.

3. From the Or select a category list box, select Statistical. Click your mouse arrow on the down arrow of the Select a function scroll bar. Select NORMINV. Click OK.

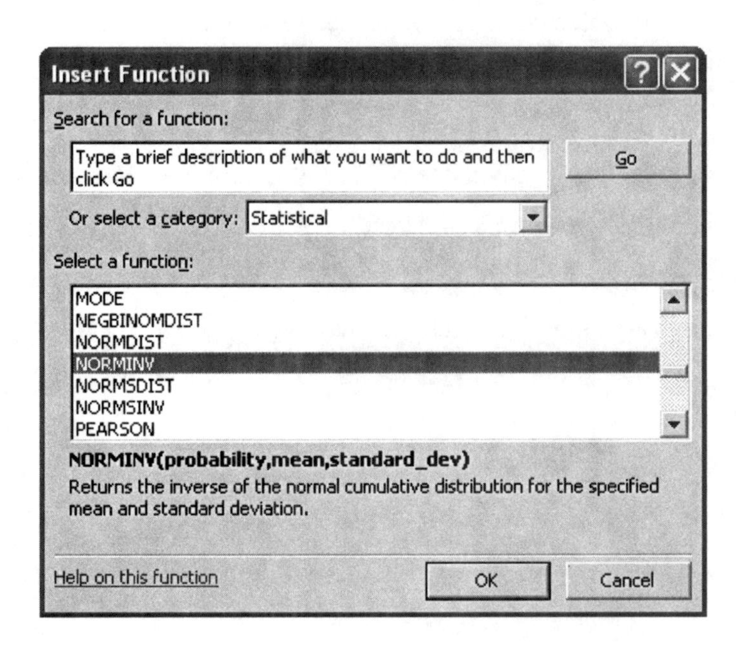

The dialog box for NORMINV is similar to the one for NORMDIST.

4. The cursor should be on the Probability text box. Key **.90**. Press the tab key.

5. In the Mean text box, key **200**. Press the tab key.

6. In the Standard_dev text box, key **20**. Click on OK.

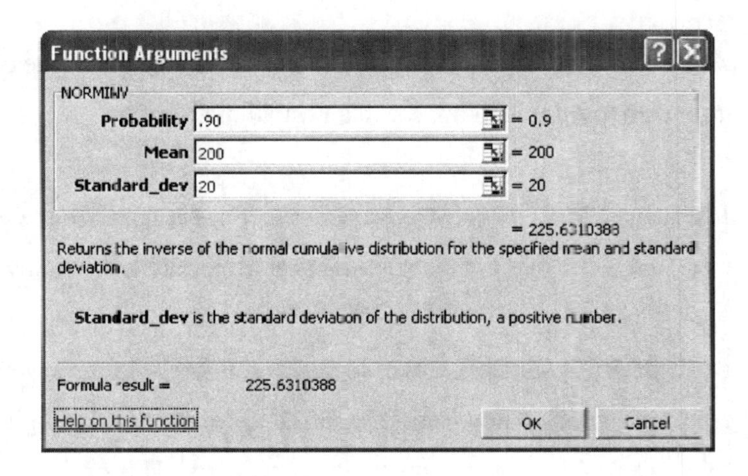

7. Use the Decrease Decimal icon to round cell D11 contents to a whole number.

8. Bold the contents of cells D10 & D11.

The result of 226 is displayed. Therefore a test score of 226 or higher will qualify. The problems for chapter 6 should show as follows. **Please Note:** Answers may vary slightly because of rounding.

	A	B	C	D	E	F	G	H
1	**Prob 1**			**Prob 2a**				
2								
3	Probability proportion			Probability proportion, area between 2 x's				
4	Area less than x				1st x		2nd x	
5								
6	Probability			CumDist=	0.308538	CumDist=	0.933193	
7	CumDist=	**78.81%**		Probability	**62.47%**			
8								
9								
10	**Prob 2b**			**Prob 3**				
11				**226**				
12	Probability proportion							
13	Area greater than x							
14								
15	CumDist=	0.864334						
16	Probability	**13.57%**						
17								

You can do several problems on one worksheet by copying cell contents. Just be sure to label your problems. It is also very helpful to bold the cell contents of the exercise number and the corresponding answer. If you wish, save your file as **Ch6probs**. Close your file.

Central Limit Theorem

The central limit theorem states that as the sample size increases, the shape of the distribution of the sample means taken from the population will approach a normal distribution.

In this exercise you will be using Excel® to select a large number of samples from a uniform population, compute the sample means, and show that the distribution is approximately normally distributed.

Example 6-2. A class of 32 statistics students was assigned to randomly select seven telephone numbers from a phone book. They were then asked to record the last digit of each phone number and find the mean of the seven numbers they recorded. This problem could illustrate the results of sampling from a uniform population if they created a histogram of the 32 sample means. It may look like the following example.

1. On a new worksheet, enter the data as shown.

	A	B	C	D	E	F	G	H	I	J	K
1	Sample	Data							Sample		
2	Number	1	2	3	4	5	6	7	Mean	Interval	Freq
3	1										

2. Highlight A1:K1.

3. From the Home tab, select Format. Select Column Width. In the Column Width text box, key 7. Click OK.

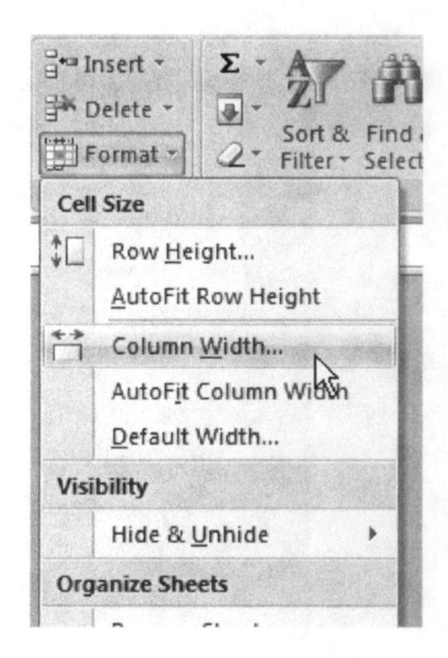

This decreases the width of the cells making it easier to view.

3. Make A3 your active cell.(The number 1 should already have been keyed) Under the Home tab, select Fill. Select Series.

4. From the Series dialog box, select Series in Columns. Select Type Linear. In the Step Value text box, key **1**. In the Stop Value text box, key **32**. Click OK.

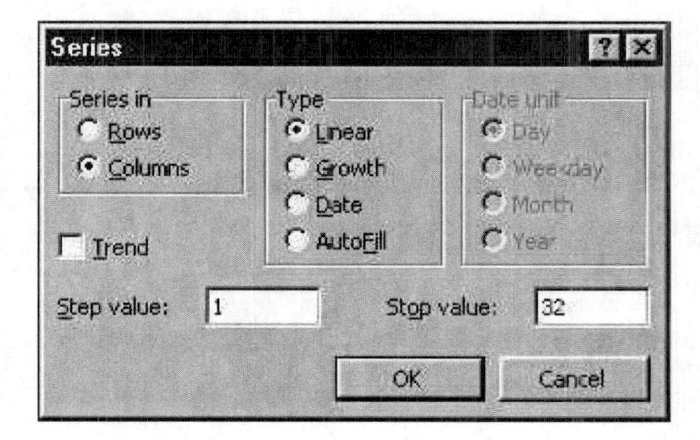

5. Make B3 your active cell. Key =**RAND()*9** (To create (), use the left and right parentheses)

This generates random numbers between 0 and 9.

6. Make B3 your active cell. Place your mouse arrow on the lower right handle of B3. Make sure you have a thick, black plus sign. Click your left mouse button twice, rapidly.

7. Cells B3:B34 should already be highlighted. Place your mouse arrow on the lower right handle of B34. Drag B34 to C34:H34.

8. While B3:H34 is still highlighted, from the Home Tab, select the Decrease Decimal icon several times until all the figures appear as whole numbers.

Random numbers between 0 and 9 have been generated in cells B3:H34.

9. In I3, key =**AVERAGE(B3:H3)**

10. Make I3 your active cell. Place your mouse arrow on the lower right handle of I3. Make sure you have a thick, black plus sign. Click your left mouse button twice, rapidly.

11. In J3:J4, key **2** and **2.4**, respectively.

12. Highlight **J3:J4**. Place your mouse arrow on the lower right handle of J4, Drag to J5:J18.

This creates interval classes in increments of .4 between 2 and 8.

13. Highlight K3:K18. With the range still highlighted, key =**FREQUENCY(I3:I34,J3:J18)**
    DO NOT PRESS THE <ENTER> KEY YET!

14. After you have finished keying, hold down the <Shift> key and the <Ctrl> key together and at the same time Press the <Enter> key. The formula in the formula bar at the top of the worksheet should be inside curly brackets, { }.

This is called an array. An array links the data together and prevents the formula from being accidentally over-written.

The first few lines of your worksheet should be similar to the one below. The numbers will be different but the format should be the same.

K3 | f_x {=FREQUENCY(I3:I34,J3:J18)}

	A	B	C	D	E	F	G	H	I	J	K
1	Sample	Data							Sample		
2	Number	1	2	3	4	5	6	7	Mean	Interval	Freq
3	1	7	3	3	4	3	1	2	3	2	0
4	2	4	6	6	3	5	0	3	4	2.4	0
5	3	6	5	3	4	6	6	5	5	2.8	0
6	4	4	8	8	1	7	0	1	4	3.2	3
7	5	9	8	8	4	6	4	3	6	3.6	2
8	6	7	6	5	2	4	8	4	5	4	1
9	7	5	4	4	3	3	2	3	4	4.4	4
10	8	3	4	1	2	9	5	8	4	4.8	2
11	9	7	2	3	6	8	4	8	5	5.2	7
12	10	6	7	7	8	6	1	1	5	5.6	8
13	11	7	4	5	1	2	7	7	4	6	4
14	12	3	9	1	6	4	1	6	4	6.4	1
15	13	8	7	6	1	4	8	5	6	6.8	0
16	14	9	3	7	2	0	4	3	4	7.2	0
17	15	0	8	7	8	5	4	2	5	7.6	0
18	16	6	9	7	4	9	1	7	6	8	0
19	17	6	7	4	8	3	5	3	5		
20	18	7	3	7	8	6	1	8	6		

Press the F9 function key on your keyboard. The generated numbers and the numbers in the frequency column should change. You will now plot the frequency distribution on a histogram. If needed, refer to Chapter 2 for additional instructions on creating a histogram.

1. From the Insert tab, select Column chart. Choose the upper left column chart. Click Next.
2. Under the Design tab, click Select Data. The Select Data Source dialog box will appear. Select Add under Legend Entry (Series).
3. In the Edit Series dialog box:
    a. Click the icon to the right of the Series values text box. Put the cursor on cell K3, click hold and drag to cell K18.
    b. Click OK.
4. Click Add under Horizontal (Category) Axis Labels.
5. In the Axis Labels dialog box, click the icon to the right of the text box Put the cursor on cell J3, click, hold and drag to cell J18. Press Enter.
6. Click OK.
7. Under the Layout tab select Axis Titles.

a. Select Primary Horizontal Axis Title and Title Below Axis. Key **Interval Weights** and press Enter.
b. Select Primary Vertical Axis Title and Rotated Axis. Key **Frequency** and press Enter.

8. Select Chart Title and Above Chart. Key **Uniform Sample**. Press Enter.
9. Select the legend on the chart and press the delete key.

A condensed chart will be displayed.

10. Make sure the frame shows around the chart. With your **right** mouse arrow, click on one of the columns. Select Format Data Series.
11. The Series Options tab should be selected. In the Gap width text box, drag on the arrow until the Gap width reads 0. Click Close.
12. With the frame still showing around the chart, click and hold the left mouse button inside the chart. A 4-way arrow will show in the chart. As you move the chart it will show as an open box. With your mouse button still depressed, drag your mouse and move your chart so the left edge of the chart is in column L and the top edge of the chart is in row 3.
13. Click on the bottom handle of the chart. Drag the bottom line to row 17.

Press the F9 function key on your keyboard and compare with the two sample charts shown above. The histogram will change according to the sample means selected. You will have a variety of charts but they should all approximate a normal curve because of the central limits theorem.

1. If you wish to show the results of this exercise, you can print only the first page of your worksheet by doing the following.
2. Highlight A1:G1. From the Home tab, select Format.
3. Select Hide & Unhide Columns. Select Hide Column. This hides part of your worksheet so you can print more easily. Click the Office button, select Print. From the Print dialog box, select Page(s).

4. In the From text box, key **1**. In the To text box, key **1**. Click OK.

If your wish, save your file as **Ch6-examp2**. Close your file.

Practice Exercises taken from textbook.

6-1. The average salary for first-year teachers is $27,989. If the distribution is approximately normal with σ = $3250, what is the probability that a randomly selected first-year teacher makes these salaries? (Textbook Exercise 6-4, Problem 2)

a. Between $20.000 and $30,000 a year

b. Less than $20,000 a year

6-2. The average age of CEOs is 56 years. Assume the variable is normally distributed. If the standard deviation is four years, find the probability that the age of a randomly selected CEO will be in the following range. (Textbook Exercise 6-4, Problem 6)

a. Between 53 and 59 years old

b. Between 58 and 63 years old

c. Between 50 and 55 years old

6-3. Full-time Ph.D. students receive an average of $12,837 per year. If the average salaries are normally distributed with a standard deviation of $1500, find these probabilities.
(Textbook Exercise 6-4, Problem 8)

a. The student makes more than $15,000.

b. The student makes between $13,000 and $14,000.

6-4. During October, the average temperature of Whitman Lake is 53.2^0 and the standard deviation is 2.3^0. Assume the variable is normally distributed. For a randomly selected day in October, find the probability that the temperature will be as follows. (Textbook Exercise 6-4, Problem 14)

a. Above 54^0

b. Below 60^0

c. Between 49^0 and 55^0

6-5. A local medical research association proposes to sponsor a footrace. The average time it takes to run the course is 45.8 minutes with a standard deviation of 3.6 minutes. If the association decides to include only the top 25% of the racers, what should be the cutoff time in the tryout run? (Textbook Exercise 6-4, Problem 16)

6-6. If the average price of a new home is $145,500, find the maximum and minimum prices of the houses a contractor will build to include the middle 80% of the market. Assume that the standard deviation of prices is $1500 and the variable is normally distributed. (Textbook Exercise 6-4, Problem 20)

6-7. To help students improve their reading, a school district decides to implement a reading program. It is to be administered to the bottom 5% of the students in the district, based on a reading achievement exam. If the average score for the students in the district is 122.6, find the cutoff score that will make a student eligible for the program. The standard deviation is 18. Assume the variable is normally distributed. (Textbook Exercise 6-4, Problem 22)

6-8. The average weight of an airline passenger's suitcase is 45 pounds. The standard deviation is 2 pounds. If 15% of the suitcases are overweight, find the maximum weight allowed by the airline. Assume the variable is normally distributed. (Textbook Chapter 6, Review Exercise # 8)

CHAPTER 7
CONFIDENCE INTERVALS AND SAMPLE SIZE

Populations are often too large to study in their entirety. Their size requires that we select samples which are then used to draw inferences about the populations. The most common estimators used to make inferences about a population are point estimates and interval estimates. A point estimate such as the mean, median, or mode uses a statistic to estimate the parameter at a single value or point. An interval estimate specifies a range within which the unknown parameter may fall. Such an interval is often accompanied by a statement as to the level of confidence placed on its accuracy. Interval estimates or confidence intervals state the range within which a population parameter probably lies.

Sample size plays an important role in determining the probability of error as well as the precision of the estimate. It is essential that the sample is representative of the target population, that it does not contain bias and that it can be replicated. The appropriate sample size is determined by the amount of dispersion in the population and the accuracy desired. It is not determined as a percent of the population. After completing this chapter, you will be able to:

- Use Excel® to calculate confidence intervals for means.
- Use Excel® to determine sample sizes for means.
- Use Excel® to calculate confidence intervals for proportions.
- Use Excel® to determine sample sizes for proportions.

Confidence Intervals for a Population Mean

The formula for the confidence interval for a population mean is $\overline{X} \pm z_{\alpha/2}\left(\frac{\sigma}{\sqrt{n}}\right)$, where $\overline{X}$ is the point estimate of the population mean, σ is the sample standard deviation, and n is the number of samples.

The $z_{\alpha/2}$ *value* depends on the level of confidence required, also called the *z*-value.

- a 99 percent confidence results in a z-value of 2.58.
- a 95 percent confidence results in a z-value of 1.96.
- a 90 percent confidence results in a z-value of 1.65.

Excel® has a statistical function called CONFIDENCE, which will compute the z confidence interval around the mean. After Excel® computes this value you need only subtract and add the value to the mean to find the confidence interval.

Problem 7-1. A researcher wishes to estimate the average amount of money a person spends on lottery tickets each month. A sample of 50 people who play the lottery found the mean to be $19 and the standard deviation to be 6.8. Find the 95% confidence interval of the population mean.

You will create a template to solve the problem.

1. On a new worksheet enter the data as shown.

	A	B	C
1	Confidence Interval-Mean		
2			
3	Mean =		
4	Interval =		
5	Confidence Limits		
6	Lower	Upper	
7			

2. In A7, key **=B3-B4**

3. In B7, key **=B3+B4**

The cell contents for A7 and B7 display 0. This is because you have not yet entered the variables.

	A	B	C
1	Confidence Interval-Mean		
2			
3	Mean =		
4	Interval =		
5	Confidence Limits		
6	Lower	Upper	
7	0.0	0.0	
8			

4. Save your file as **cnfdinv-mean.**

All you need to do now is enter the variable numbers.

5. In B3, enter **19**, which is the mean of the problem.

Make cell B4 your active cell.

6. From the Formula bar click on the Insert Function icon. 

The Insert Function dialog box is displayed

7. From the Or select a category scroll bar, select Statistical.

8. From the Select a function scroll bar, select CONFIDENCE. Click on OK.

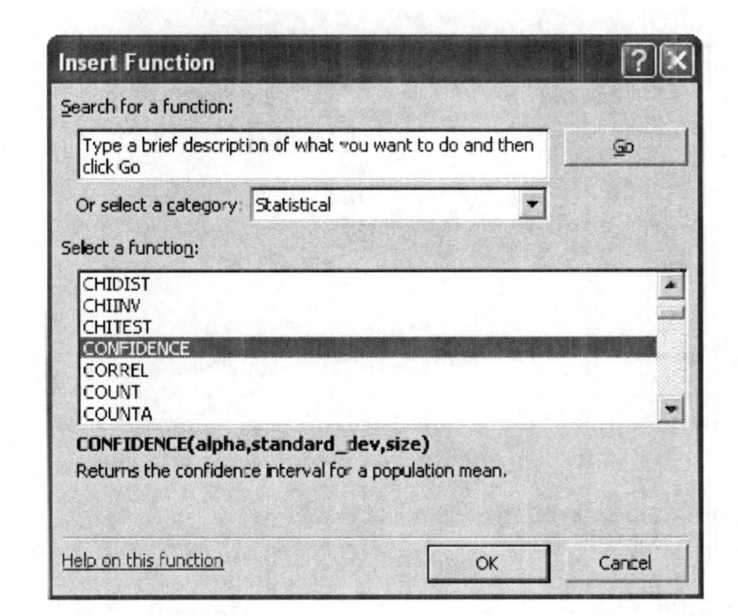

The CONFIDENCE dialog box contains several text boxes to fill.

- The *Alpha* text box is for the significance level used to compute the confidence level. It will be the complement of the given desired confidence level.
- The *Standard_dev* text box is for the population standard deviation for the data range, σ.
- The *Size* text box is for the sample size, n.

You will use this dialog box to complete the problem.

9. Your cursor should be on the Alpha text box. Key **.05** since that is the complement of the 95% desired confidence level. Press the tab key.

10. Your cursor should be on the Standard_dev text box. Key **6.8.** Press the tab key.

11. Your cursor should be on the Size textbox, key **50**.

12. Click OK.

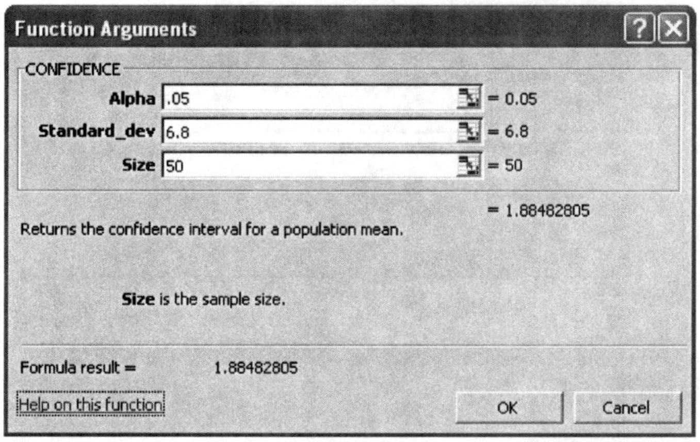

Function Arguments

CONFIDENCE

Alpha .05 = 0.05

Standard_dev 6.8 = 6.8

Size 50 = 50

= 1.88482805

Returns the confidence interval for a population mean.

Size is the sample size.

Formula result = 1.88482805

Help on this function

OK Cancel

The lower and upper limits are displayed in cells A7 and B7.

13. Highlight A7:B7. Under the Home tab click on the Decrease Decimal icon to show the values to one decimal place.

14. Bold cell A7:B7.

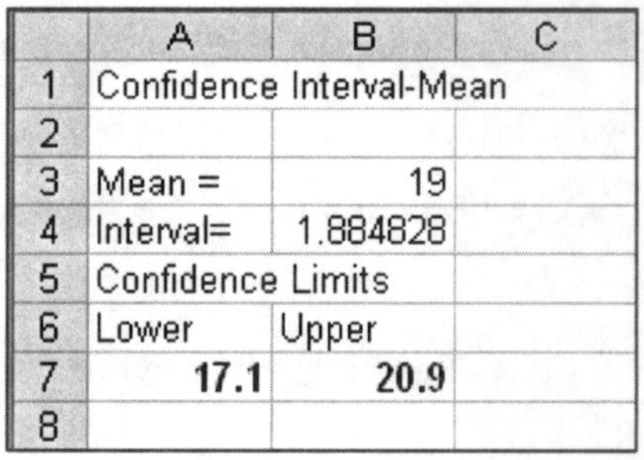

	A	B	C
1	Confidence Interval-Mean		
2			
3	Mean =	19	
4	Interval=	1.884828	
5	Confidence Limits		
6	Lower	Upper	
7	**17.1**	**20.9**	
8			

Save your file as **Ch7-prob1**

You can use the **cnfdinv-mean** template to solve other problems finding the confidence interval of the mean.

Determining a sample size for means.

The formula for determining a sample size for a mean is $n = \left(\frac{z_{\alpha/2} \cdot \sigma}{E} \right)^2$ where σ is the estimate of the population standard deviation, and *E* is the maximum allowable error.

The $z_{\alpha/2}$ *value* depends on the level of confidence required.

- A 99 percent confidence results in a z-value of 2.58.
- A 95 percent confidence results in a z-value of 1.96.

- A 90 percent confidence results in a z-value of 1.65.

Problem 7-2. The college president asks the statistics teacher to estimate the average age of the students at their college. How large a sample is necessary? The statistics teacher would like to be 99% confident that the estimate should be accurate within 1 year. From a previous study, the standard deviation of the ages is known to be 3 years.

Excel® does not have any function for this so you need to create a template to solve the formula.

1. On a new worksheet, in A1, key **Sample size for mean**.

2. In A3, key **z-value=**

3. In A4, key **StdDev=**

4. In A5, key **error=**

5. In A6 key **SaSize=**

6. In A7, key **rounded=**

B7 | fx =INT(B6+0.99)

	A	B	C	D
1	Sample size for mean			
2				
3	z-value=			
4	StdDev=			
5	error=			
6	SaSize=	#DIV/0!		
7	roun=	#DIV/0!		
8				

7. In B6, key **=((B3*B4)/B5)^2**

8. In B7, key **=INT(B6+.99)** This rounds up the answer to the next whole number.

The contents of B6 and B7 will display #DIV/0!. They will change when you fill in the values.

9. Save your file as **sample size mean**

You will now fill in the values to solve the problem.

10. In B3, key **2.58**, since the confidence required is 99 percent.

11. In B4, key **3**, the estimated standard deviation.

12. In B5, key **1**, since the error is to be less than 1.

The sample size of 59.9076 is displayed in B6.

The rounded value of 60 is displayed in B7.

13. Bold the contents of A1 and B7.

Save you file as **Ch7-prob2**

	A	B	C
1	**Sample size for mean**		
2			
3	z-value=	2.58	
4	StdDev=	3	
5	error=	1	
6	SaSize=	59.9076	
7	rounded=	**60**	
8			

Confidence Interval for a Population Proportion

The formula for determining the confidence interval for a proportion is $\hat{p} \pm (z_{\alpha/2})\sqrt{\frac{\hat{p}\hat{q}}{n}}$ where: $\hat{p}$ is the sample proportion, n is the sample size, and $\hat{q}$ is 1- $\hat{p}$.

The $z_{\alpha/2}$ *value* depends on the level of confidence required.

- a 99 percent confidence results in a z-value of 2.58.
- a 95 percent confidence results in a z-value of 1.96.
- a 90 percent confidence results in a z-value of 1.65.

Problem 7-3. A sample of 500 nursing applications included 60 from men. Find the 90% confidence interval of the true proportion of men who applied to the nursing program.

We have to create a template to solve the formula.

1. On a new worksheet enter the data as shown.

2. In B6, key **=B4*SQRT(B3*(1-B3)/B5)**

	A	B	C
1	Confidence Interval-Proportion		
2			
3	Propor=		
4	z=		
5	n=		
6	Interval=		
7	Confidence Limits		
8	Lower	Upper	
9			

3. In A9, key =**B3-B6***

4. In B9, key =**B3+B6**

The cell contents for B6, A9 and B9 display #DIV/0! This is because you have not yet entered the variables.

	A	B	C
1	Confidence Interval-Proportion		
2			
3	Propor=		
4	z=		
5	n=		
6	Inter	#DIV/0!	
7	Confidence Limits		
8	Lower	Upper	
9	#DIV/0!	#DIV/0!	
10			

5. Save your file as **cnfdinv-pro**

All you need to do now is enter the variable numbers.

6. In B3, key **=60/500,** this is the proportion of the sample.

7. In B4, key **1.65**, since the confidence level is 90 percent.

8. In B5, key **500**, the sample size.

The lower and upper limits are displayed in A9 and B9.

9. Highlight A9 and B9.

10. In the Tool bar, click on the Percent Style icon. %

11. In the Tool bar, to change to one place past the decimal click on the Increase Decimal icon.

12. Bold cells A9 and B9.

	A	B	C
1	Confidence Interval-Proportion		
2			
3	Propor=	0.12	
4	z=	1.65	
5	n=	500	
6	Interval=	0.023979	
7	Confidence Limits		
8	Lower	Upper	
9	**9.6%**	**14.4%**	
10			

If you wish, save your worksheet as **Ch7-prob3**.
Close your file.

Determining a sample size for proportion.

The formula for determining a sample size for a proportion is $n = \hat{p}\hat{q}\left(\frac{z_{\alpha/2}}{E}\right)^2$ where:

$\hat{p}$ is the sample proportion. If the proportion is not known, $\hat{p}$ is assigned a value of .5.

$\hat{q}$ is 1- $\hat{p}$

E is the margin of error in the proportion that is requested.

The $z_{\alpha/2}$ *value* depends on the level of confidence required.

- 99 percent confidence results in a z-value of 2.58.
- 95 percent confidence results in a z-value of 1.96.
- 90 percent confidence results in a z-value of 1.65.

Problem 7-4 A researcher wishes to estimate, with 95% confidence, the proportion of people who own a home computer. A previous study shows that 40% of those interviewed had a computer at home. The researcher wishes to be accurate within 2% of the true proportion. Find the minimum sample size necessary.

We have to create a template to solve the formula.

1. On a new worksheet, in cell A1, key **Sample size for proportion.**

2. In A3, key **z-value**=

3. In A4, key **SaPro=**

4. In A5, key **error=**

5. In A6, key **SaSize=**

6. In A7 , key **rounded=**

7. In B6, key =**B4*(1-B4)*(B3/B5)^2**

8. In B7, key **=INT(B6+.99)** This rounds up the answer to the next whole number.

The contents of B6 and B7 will display #DIV/0!. They will change when you fill in the values.

B7		f_x =INT(B6+0.99)	
	A	B	C	D
1	Sample size for proportion			
2				
3	z-value=			
4	SaPro=			
5	error=			
6	SaSize=	#DIV/0!		
7	round=	#DIV/0!		
8				

9. Save your file as **sample size pro**

10. In B3, key **1.96**, since the confidence required is 95 percent.

11. In B4, key **.40**, the population proportion.

12. In B5, key **.02**, since you want to be accurate within 2%.

The sample size of 2304.96 is displayed in B6. The rounded value is displayed in B7.

13. Bold the contents of A1 and B7.

	A	B	C
1	**Sample size for proportion**		
2			
3	z-value=	1.96	
4	SaPro=	0.4	
5	error=	0.02	
6	SaSize=	2304.96	
7	rounded=	**2305**	
8			

Save your worksheet as **Ch7-prob4**.

Inserting Templates on a Worksheet

You can do several problems on a worksheet. Just copy the cells containing the templates from one worksheet to another worksheet. For example if you wanted to use all four templates you created for this chapter, you would do the following:

1. Open the files, **cnfdinv-mean**, **cnfdinv-pro**, **sample size mean** and **sample size pro**.

2. Open a new workbook.

3. In cell C1 key, **Confidence Intervals and Sample Size**

4. Under the View tab, click Switch Windows. Select cnfdinv-mean.

You are now in the cnfdinv-mean workbook.

5. Highlight A1:B7.

6. Under the Home tab, select the Copy icon.

7. Under the View tab, click Switch Windows. Select Book 2 (or whatever book the new worksheet is).

8. Make A3 your active cell.

9. Under the Home tab, click your mouse arrow on the Paste icon, and press the <Enter> key.

10. Under the View tab, click Switch Windows. Select cnfdinv-pro.

You are now in the cnfdinv-pro worksheet.

11. Highlight A1:B9.

12. Under the Home tab, select the Copy icon.

13. Under the View tab, click Switch Windows. Select Book 2 (or whatever book the new worksheet is).

14. Make E3 your active cell.

15. From the Home tab click your mouse arrow on the Paste icon, and press the <Enter> key.

16. Under the View tab, click Switch Windows. Select sample size mean.

You are now in the sample size mean worksheet.

17. Highlight A1:B7.

18. From the Home tab select the Copy icon.

19. Under the View tab, click Switch Windows. Select Book 2 (or whatever Book the new worksheet is).

20. Make A14 your active cell.

21. From the Home tab click your mouse arrow on the Paste icon, and press the <Enter> key.

22. Under the View tab, click Switch Windows. Select sample size pro.

You are now in the sample size pro worksheet.

23. Highlight A1:B7.

24. From the Home tab select the Copy icon.

25. Under the View tab, click Switch Windows. Select Book 2 (or whatever book the new worksheet is).
26. Make E14 your active cell.

27. From the Home tab click your mouse arrow on the Paste icon, and press the <Enter> key.

	A	B	C	D	E	F	G
1			Confidence Intervals and Sample Size				
2							
3	Confidence Interval-Mean				Confidence Interval-Proportion		
4							
5	Mean =				Propor=		
6	Interval=				z=		
7	Confidence Limits				n=		
8	Lower	Upper			Interval=	#DIV/0!	
9	0	0			Confidence Limits		
10					Lower	Upper	
11					#DIV/0!	#DIV/0!	
12							
13							
14	Sample size for mean				Sample size for proportion		
15							
16	z-value=				z-value=		
17	StdDev=				SaPro=		
18	error=				error=		
19	SaSize=	#DIV/0!			SaSize=	#DIV/0!	
20	rounded=	#DIV/0!			rounded=	#DIV/0!	
21							

All four templates are on one worksheet. As you use each template, make sure the appropriate cell is active before you key in the value.

Be sure to identify each problem and bold the problem numbers and answers to distinguish them more easily.

Save your file as **Chapter 7 templates**. Close all your files.

Practice Exercises taken from textbook.

7-1. A study of 40 English composition professors showed that they spent, on average, 12.6 minutes correcting a student's term paper. Find the 90% confidence interval of the mean time for all composition papers when σ = 2.5 minutes. (Textbook Exercise 7-2, Problem 13 b)

7-2. A study of 36 marathon runners showed that they could run at an average rate of 7.8 miles per hour. The sample standard deviation is 0.6. Find the 90% confidence interval for the mean of all runners. (Textbook Chapter 7, Review Exercise # 1)

7-3. A pizza shop owner wishes to find the 95% confidence interval of the true mean cost of a large plain pizza. How large should the sample be if she wishes to be accurate to within \$0.15? A previous study showed that the standard deviation of the price was \$0.26. (Textbook Exercise 7-2, Problem 24)

7-4. A researcher is interested in estimating the average salary of teachers in a large urban school district. She wants to be 95% confident that her estimate is correct. If the standard deviation is \$1050, how large a sample is needed to be accurate to within \$200? (Textbook Chapter 7, Review Exercise # 6)

7-5. A survey of 50 first-time white-water canoers showed that 23 did not want to repeat the experience. Find the 90% confidence interval of the true proportion of canoers who did not wish to canoe the rapids a second time. (Textbook Exercise 7-4, Problem 10)

7-6. A U.S. Travel Data Center's survey of 1500 adults found that 42% of respondents stated that they favor historical sites as vacations. Find the 95% confidence interval of the true proportion of all adults who favor visiting historical sites as vacations. (Textbook Chapter 7, Review Exercise # 8)

7-7. A recent study indicated that 29% of the 100 women over age 55 in the study were widows. (Textbook Exercise 7-4, Problem 16)

a. How large a sample must one take to be 90% confident that the estimate is within 0.05 of the true proportion of women over age 55 who are widows?

b. If no estimate of the sample proportion is available, how large should the sample be?

7-8. A federal report indicated that 27% of children ages 2 to 5 years had a good diet – an increase over previous years. How large a sample is needed to estimate the true proportion of children with good diets within 2% with 95% confidence? (Textbook Exercise 7-4, Problem 20)

CHAPTER 8
HYPOTHESIS TESTING

Chapter 7 dealt with a segment of statistical inference called estimation. This chapter will deal with a method of testing those estimations. Hypothesis testing is a procedure based on sample evidence and probability theory. It is used to determine whether the hypothesis is a reasonable statement and should not be rejected, or is unreasonable and should be rejected. When we do this we also evaluate the risks involved in making these decisions based on sample information and the interrelationship of these risks based on sample size. The use of Excel® will simplify this process.

Using Excel®, we will analyze the differences between the results actually observed and the results we would expect to obtain if some underlying hypothesis were actually true. This chapter will illustrate how to:

- Use the z Test to test means of large samples, and population proportions
- Use the t Test to test means of small samples
- Use the chi-square test to test variances

Steps in hypothesis testing:

These five steps in hypothesis testing will help you solve a wide variety of hypothesis problems, not just those using a z test for large samples. The five-step process provides a similar format and a thread of continuity that can be helpful in recognizing different statistical tests and knowing which test to apply different situations.

1. **State the null and alternative hypothesis** using either formulas or words. The Null Hypothesis (H_o) is always the statement of no significant difference. The Alternative Hypothesis (H_1) is always the statement that there is a significant difference. When direction is stated it is a one-directional test (one-tailed). When direction is not stated it is a two-directional test (two-tailed).

2. **State the level of significance** or the probability that the null hypothesis is rejected when, in fact, it is true.

3. **State the statistical test** you will be using: the z test, t test, f test, chi square test, etc.

4. **Formulate a decision rule**. Using a picture or curve that estimates the distribution you are testing, show the critical value if you are performing a one-directional test or the upper and lower critical values if you are performing a two-directional test.

5. **Do it**. Show the formula you used and at least the major steps involved. State the results of the hypothesis test in terms of the question using complete sentences and examples.

Creating Names

In Chapter 8 you will be using formulas to solve problems. You will also need to build an Excel® worksheet before completing a hypothesis test. In building this worksheet, it is often helpful to use names in a formula instead of cell references. The following exercise will give you some experience naming cells and using the names in a formula: $\mu = \frac{\Sigma X}{N}$

Example 8-1. There are 6 students in a computer class. Their test scores were 92, 96, 61, 86, 79, and 84. What is the mean test grade?

The formula for finding the population mean is the sum of all the values in the population divided by the number of values in the population.

1. On a new worksheet, enter the following as shown.

	A	B	C	D
1		Student	Score	
2		(N)	(X)	
3		A	92	
4		B	96	
5		C	61	
6		D	86	
7		E	79	
8		F	84	
9	Sum_of_N	6		Sum_of_X
10				
11		mean		
12				

2. Highlight B3:B8. From the Home Tab, click on the Align Right icon.

3. Highlight C3:C8. Under the Home Tab, click on the Sum icon.

The sum of X (498) is displayed in cell C9.

4. Highlight A9:B9. Under the Formulas tab, select Create from Selection. Select the check box for Left Column. Click OK.

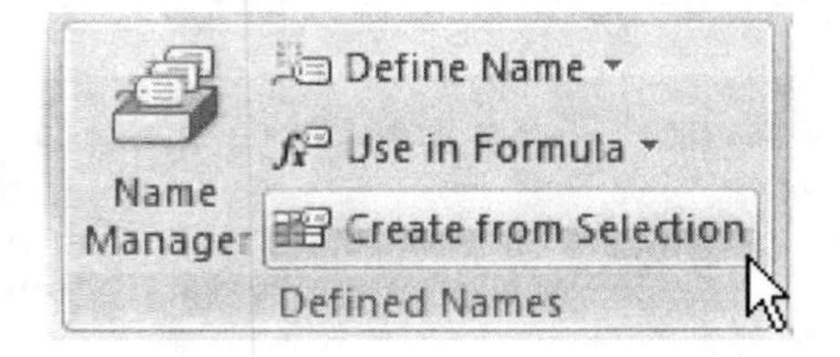

Notice that when B9 is selected as the active cell, the name of the cell (Sum_of_N) shows in the Name box in the upper left corner.

5. Highlight C9:D9. Under the Formulas tab, select Create from Selection. Select the check box for Right Column. Click on OK.

Notice that when C9 is selected as the active cell, the name of the cell (Sum_of_X) shows in the Name box in the upper left corner.

6. Make C11 your active cell. Key **=Sum_of_X/Sum_of_N**. Press <Enter>. The population mean (83) is displayed.

You could have keyed the formula =C9/B9, but you used the names of those cells instead.

Notice that when C11 is selected as the active cell, the formula =*Sum_of_X/Sum_of_N* shows in the formula bar at the top of the worksheet.

The previous exercise was to prepare you for creating formulas using names of cells, which is often easier to visualize than using just the cell references.

C11 | fx =Sum_of_X/Sum_of_N

	A	B	C	D	E
1		Student	Score		
2		(N)	92		
3		A	96		
4		B	61		
5		C	86		
6		D	79		
7		E	84		
8		F			
9	Sum_of_N	6	498	Sum_of_X	
10					
11		mean	83		
12					

You will create a worksheet for hypothesis testing of large samples. It can be used for one left-tail, one right-tail, or a two-tailed hypothesis test. After the worksheet is created, you will be able to enter the input data. The worksheet will compute the results for all three kinds of hypothesis tests. You must then decide which one is needed for your test.

z Test for a Mean. The z test is a test of hypothesis about one sample mean from a large population. The worksheet you create will include formulas used in your textbook. The symbols will often be referred to as names. The formula for finding the test statistic z is $z = \dfrac{\overline{X} - \mu}{\sigma / \sqrt{n}}$, where

$\overline{X}$ is the sample mean. It will be referred to as SaMean.

μ is the hypothesized population mean. It will be referred to as HoMean.

σ is the standard deviation of the population sample. It will be referred to as StdDev.

n is the sample number size. It will be referred to as n.

Example 8-2. Creating a worksheet for testing of normal hypothesis of a one sample (large population) mean.

	A	B	C
1	Test of Normal Hypotheses:		
2	One Sample Mean-Large		
3			
4	Input Data		
5	HoMean		
6	SaMean		
7	n		
8	StdDev		
9	Alpha		
10	Calculated Value		
11	z		
12	Test for Left-Tail		
13	LftCrt_zVal		
14	Conclusion		
15	p-value		
16	Test for Right-Tail		
17	RtCrt_zVal		
18	Conclusion		
19	p-value		
20	Test for Two-Tail		
21	AbsCrt_zVal		
22	Conclusion		
23	p-value		
24			

1. On a new worksheet, enter the following data as shown.

2. Widen column A by double clicking your mouse arrow between column headings A and B.

3. Highlight A5:A9. Under the Home Tab, click on the Align Right icon. 

4. Right Align cell A11. Right align Cells A13:A15. Right align Cells A17:A19. Right align cells A21:A23.

5. Highlight A5:B9. Under the Formulas tab, select Create from Selection. Select the check box for Left Column. Click on OK. The cells B5:B9 have names as well as cell references.

6. Highlight A11:B11. Use the same method from Step 5 to create names.

7. Use step 5 to create a name for A13:B13, A17:B17, and A21:B21

8. Make B11 your active cell. Key =**(SaMean-HoMean)/(StdDev/SQRT(n))** Press <Enter>.

This is the formula for finding z.

The output result in cell B11 shows #DIV/0!. This is because none of the input cells have values yet (so the formula is trying to divide by 0). When you start entering the input data, the formulas will generate the correct output.

9. Make B13 your active cell. Key =**NORMSINV(Alpha)** Press <Enter>.

This is the formula for finding the left critical value of z. There is no value for the output because there has been no input data.

	A
1	Test of Normal Hypotheses:
2	One Sample Mean-Large
3	
4	Input Data
5	HoMean
6	SaMean
7	n
8	StdDev
9	Alpha
10	Calculated Value
11	z
12	Test for Left-Tail
13	LftCrt_zVal
14	Conclusion
15	p-value
16	Test for Right-Tail
17	RtCrt_zVal
18	Conclusion
19	p-value
20	Test for Two-Tail
21	AbsCrt_zVal
22	Conclusion
23	p-value
24	

As you key the remaining formulas, there will be no output value for any of the cells. The values will be obtained when the input data is entered later.

After each formula is completed, press the <Enter> key. If you make a mistake and have entered the formula incorrectly, double click your mouse arrow on the cell and you will be able to edit the cell without re-keying the entire contents.

10. Make B17 your active cell. Key =**-NORMSINV(Alpha)**

This is the formula for finding the right critical value of z.

11. Make B21 your active cell. Key =**ABS(NORMSINV(Alpha/2))**

This is the formula for finding the critical value of z on a two-tailed test.

12. Make B14 your active cell. Key **=If(z<LftCrt_zVal, "Reject Ho","Do Not Reject Ho")**

This is the decision to reject or not to reject the null hypothesis of a left-tail test. If the test statistic of z is less than the left critical z value, then the hypothesis is rejected, otherwise do not reject the null hypothesis.

13. Make B18 your active cell. Key **=If(z>RtCrt_zVal, "Reject Ho", "Do Not Reject Ho")**

This is the decision to reject or not to reject the null hypothesis of a right-tail test. If the test statistic of z is greater than the right critical z value, then the hypothesis is rejected. Otherwise do not reject the null hypothesis.

14. Make B22 your active cell. Key **=If(OR(z<-AbsCrt_zVal, z>AbsCrt_zVal), "Reject Ho", "Do Not Reject Ho")**

This is a decision to reject or not to reject the null hypothesis of a two-tailed test. If the test statistic of z is less than the negative critical z value, *or* if the test statistic of z is greater than the critical z value, then reject the hypothesis. Otherwise do not reject the hypothesis. You now need to enter the formula for the p-value for each of the three tests.

15. Make B15 your active cell. Key **=NORMSDIST(z)**

16. Make B19 your active cell. Key **=1-NORMSDIST(z)**

17. Make B23 your active cell. Key **=If(z>0,2*(1-NORMSDIST(z)),2*NORMSDIST(z))**

Remember: To display the formula view (instead of the output), hold down the <Ctrl> key and tap the <~> key (usually located above the tab key). Use this same key combination to toggle between views.

	A	B	C	D	E	F	G	H
1	Test of Normal Hypotheses:							
2	One Sample Mean-Large							
3								
4	Input Data							
5	HoMean							
6	SaMean							
7	n							
8	StdDev							
9	Alpha							
10	Calculated Value							
11	z	=(SaMean-HoMean)/(StdDev/SQRT(n))						
12	Test for Left-Tail							
13	LftCrt_zVal	=NORMSINV(Alpha)						
14	Conclusion	=IF(z<LftCrt_zVal, "Reject Ho", "Do Not Reject Ho")						
15	p-value	=NORMSDIST(z)						
16	Test for Right-Tail							
17	RtCrt_zVal	=-NORMSINV(Alpha)						
18	Conclusion	=IF(z>RtCrt_zVal, "Reject Ho","Do Not Reject Ho")						
19	p-value	=1-NORMSDIST(z)						
20	Test for Two-Tail							
21	AbsCrt_zVal	=ABS(NORMSINV(Alpha/2))						
22	Conclusion	=IF(OR(z<-AbsCrt_zVal,z>AbsCrt_zVal),"Reject Ho","Do Not Reject Ho")						
23	p-value	=IF(z>0,2*(1-NORMSDIST(z)),2*NORMSDIST(z))						
24								

As you are entering the formulas, your worksheet will display the cell contents as shown above.

B23 f_x =IF(z>0, 2*(1-NORMSDIST(z)),2*NORMSDIST(z))

	A	B	C	D	E	F
1	Test of Normal Hypotheses					
2	One Sample Mean-Large					
3						
4	Input Data					
5	HoMean					
6	SaMean					
7	n					
8	StdDev					
9	Alpha					
10	Calculated Value					
11	z	#DIV/0!				
12	Test for Left-Tail					
13	LftCrt_zVal	#NUM!				
14	Conclusion	#DIV/0!				
15	p-value	#DIV/0!				
16	Test for Right-Tail					
17	RtCrt_zVal	#NUM!				
18	Conclusion	#DIV/0!				
19	p-value	#DIV/0!				
20	Test for Two-Tail					
21	AbsCrt_zVal	#NUM!				
22	Conclusion	#DIV/0!				
23	p-value	#DIV/0!				
24						

Save a copy of this worksheet so you can enter data into the input cells and not have to recreate the worksheet each time. Save as **1sa-mean.** You will be retrieving this file several times and revising it to create other templates. When you make changes be sure to save the file as a different name, leaving file 1sa-mean in its original form.

Problem 8-1. A researcher reports that the average salary of assistant professors is more than $42,000. A sample of 30 assistant professors has a mean salary of $43,260. At $\alpha = 0.05$, test the claim that assistant professors earn more than $42,000 a year. The standard deviation of the population is $5230.

1. Open the file **1sa-mean** if it is not already active.

2. In cells B5:B9, enter the input data in the appropriate cells, as shown below.

	A	B
1	Test of Normal Hypotheses:	
2	One Sample Mean-Large	
3		
4	Input Data	
5	HoMean	42000
6	SaMean	43260
7	n	30
8	StdDev	5230
9	Alpha	0.05
10		

The values in the output cells have automatically changed to reflect the input data.

Bold cell B11, the z-value and cell B18, the conclusion. You can then interpret the results. This is a right-tailed test. Since the calculated value for z of 1.32 (rounded) is lower than the critical value 1.64 (rounded), we do not reject the null hypotheses.

There is not enough evidence to support the claim that assistant professors earn more than $42,000 a year. The p-value tells you that you have a .093491 chance of rejecting a true null hypothesis (a type one error).

	A	B	C
1	Test of Normal Hypotheses		
2	One Sample Mean-Large		
3			
4	Input Data		
5	HoMean	42000	
6	SaMean	43260	
7	n	30	
8	StdDev	5230	
9	Alpha	0.05	
10	Calculated Value		
11	z	**1.319561**	
12	Test for Left-Tail		
13	LftCrt_zVal	-1.64485	
14	Conclusion	Do Not Reject Ho	
15	p-value	0.906509	
16	Test for Right-Tail		
17	RtCrt_zVal	1.644853	
18	Conclusion	**Do Not Reject Ho**	
19	p-value	0.093491	
20	Test for Two-Tail		
21	AbsCrt_zVal	1.959961	
22	Conclusion	Do Not Reject Ho	
23	p-value	0.186982	
24			

If you wish, save your worksheet as **Ch8-prob1**. Close your file.

t Test for a Mean. A t test is for a population mean from a small sample.

If the sample size is small or less than 30 observations, and the standard deviation is unknown, then the z distribution previously used is not appropriate. In that case, the **student t**, or the **t distribution**, is used as the test statistic.

Earlier you created a worksheet for testing large samples. You will need to modify this worksheet to test small samples.

The formula for a population t test is: $t = \dfrac{\overline{X} - \mu}{s / \sqrt{n}}$

$\overline{X}$ is the mean of the small sample. It will be referred to as SaMean.

μ is the hypothesized population mean. It will be referred to as HoMean.

s is the standard deviation of the sample. It will be referred to as StdDev.

n is the sample size. It will be referred to as *n*.

The decision to reject or not to reject the hypothesis (the t test) is very similar to the decision using the z test.

You will rename some cells, edit the formulas of some cells, and re-key the formulas in other cells. The formula for t tests requires the use of the degrees of freedom, df (df is found by the formula df = n-1).

Example 8-3. Creating a worksheet for testing a population mean from a small sample.

1. Open **1sa-mean**. Key **Small Samples** in A1 and **Population Mean** in A2.

2. Make A11 your active cell. Under the Home tab, click the drop down arrow by Insert and select Insert Sheet Rows.

The variables in A5:A9 remain the same as for large samples, however some are computed differently.

3. Make B6 your active cell. Key **=AVERAGE(Sample)**. (Don't forget to press the <Enter> key after entering each cell's contents.)

This computes the mean of your sample data, which you will enter later.

4. Make B8 your active cell. Key **=STDEV(Sample)**

This computes the standard deviation of your sample data, which you will enter later.

5. In cell A11, key **df**

6. Right align A11.

7. Highlight A11:B11 Under the Formulas tab, select Create from Selection. Select Left Column. Click OK.

8. Make B11 your active cell. Key **=n-1**

This computes the degrees of freedom.

9. In A12, key **t**

10. Highlight A12:B12. Repeat the instructions in step 7 for creating a name.

You will notice that when B12 is your active cell, the name of the cell (t) shows in the name box in the upper left corner. B12 used to have the name z. You have renamed the cell, *t*.

11. Make A14 your active cell. Double click the left mouse button.

The cursor is blinking inside the cell. The cell can now be edited.

12. Use the left arrow key to move the flashing vertical bar to the left of the letter z. Press the <delete> key once. Key the letter **t**. (Remember to press the <Enter> key when you are finished with editing.)

Cell A14 should now read, LftCrt_tVal.

13. Make **B14** your active cell. Key = **-TINV(2*Alpha,df)**

Make sure that there is no space between the equals and the minus signs. This replaces the old formula for LftCrt_zVal with the new formula for LftCrt_tVal.

14. Highlight A14:B14. Use the instructions in step 7 for creating a name.

15. Make B15 your active cell. Double click the left mouse button to edit the cell. Using your arrow keys to move the flashing vertical bar, delete the letter z both times and replace with the letter **t**

Cell B15 should now read, =IF(t<LftCrt_tVal, "Reject Ho", "Do Not Reject Ho")

16. Make B16 your active cell. Key =**If(t<0,TDIST(ABS(t),df,1),1-TDIST(t,df,1))**

This replaces the old formula for *p* with the new formula using the t test. You will continue to edit some cell contents and re-key others.

17. Make A18 your active cell. Double click to edit the cell. Replace the letter z with the letter **t**.

Cell A18 should now read RtCrt_tVal.

18. Make B18 your active cell. Key =**TINV(2*Alpha,df)**

19. Highlight A18:B18. Use the instructions in step 7 for creating a name.

20. Make B19 your active cell. Double click to edit the cell. Delete the letter z both times and replace with the letter **t**

Cell B19 should now read, =If(t>RtCrt_tVal, "Reject Ho", "Do Not Reject Ho")

21. Make B20 your active cell. Key =**IF(t>0, TDIST(t,df,1),1-TDIST(ABS(t),df,1))**

22. Make A22 your active cell. Double click to edit the cell. Replace the letter z with the letter **t**

Cell A22 should now read, AbsCrt_tVal.

23. Make B22 your active cell. Key **=TINV(Alpha, df)**

24. Highlight A22:B22. Use the instructions in step 7 for creating a name.

25. Make B23 your active cell. Double click to edit the cell. Delete the letter z each time and replace with the letter **t**

Cell B23 should now read,

= IF(OR(t<-AbsCrt_tVal,t>AbsCrt_tVal), “Reject Ho”, “Do not Reject Ho”)

26. Make B24 your active cell. Key **=TDIST(ABS(t),df,2)**

After you have finished editing, and re-keying the worksheet, it should have the cell contents as shown.

	A	B	C	D	E	F	G	H
1	Small Samples							
2	Population Mean							
3								
4	Input Data							
5	HoMean							
6	SaMean	=AVERAGE(Sample)						
7	n							
8	StdDev	=STDEV(Sample)						
9	Alpha							
10	Calculated Value							
11	df	=n-1						
12	t	=(SaMean-HoMean)/(StdDev/SQRT(n))						
13	Test for Left-Tail							
14	LftCrt_tVal	=-TINV(2*Alpha,df)						
15	Conclusion	=IF(t<LftCrt_tVal, "Reject Ho", "Do Not Reject Ho")						
16	p-value	=IF(t<0,TDIST(ABS(t),df,1),1-TDIST(t,df,1))						
17	Test for Right-Tail							
18	RtCrt_tVal	=TINV(2*Alpha,df)						
19	Conclusion	=IF(t>RtCrt_tVal, "Reject Ho","Do Not Reject Ho")						
20	p-value	=IF(t>0,TDIST(t,df,1),1-TDIST(ABS(t),df,1))						
21	Test for Two-Tail							
22	AbsCrt_tVal	=TINV(Alpha,df)						
23	Conclusion	=IF(OR(t<-AbsCrt_tVal,t>AbsCrt_tVal),"Reject Ho","Do Not Reject Ho")						
24	p-value	=TDIST(ABS(t),df,2)						
25								

The worksheet should appear as shown.

	A	B
1	Small Samples	
2	Population Mean	
3		
4	Input Data	
5	HoMean	
6	SaMean	#NAME?
7	n	
8	StdDev	#NAME?
9	Alpha	
10	Calculated Value	
11	df	-1
12	t	#NAME?
13	Test for Left-Tail	
14	LftCrt_tVal	#NUM!
15	Conclusion	#NAME?
16	p-value	#NAME?
17	Test for Right-Tail	
18	RtCrt_tVal	#NUM!
19	Conclusion	#NAME?
20	p-value	#NAME?
21	Test for Two-Tail	
22	AbsCrt_tVal	#NUM!
23	Conclusion	#NAME?
24	p-value	#NAME?
25		

Save this worksheet as **t-tst1mn**.

Problem 8-2. An educator claims that the average salary of substitute teachers in school districts in Allegheny County, Pennsylvania, is less than \$60 per day. A random sample of eight school districts is selected, and the daily salaries (in dollars) are shown. Is there enough evidence to support the educators claim at $\alpha = 0.10$?

60, 56, 60, 55, 70, 55, 60, 55

1. Open the file **t-tst1mn**, if it is not already open.

2. Make D1 your active cell. Key **Sample**

Don't forget to press the <Enter> key after entering the contents of each cell.

3. In D2:D9, key in the sample weights as shown below.

	A	B	C	D
1	Small Samples			Sample
2	Population Mean			60
3				56
4	Input Data			60
5	HoMean			55
6	SaMean	#NAME?		70
7	n			55
8	StdDev	#NAME?		60
9	Alpha			55
10				

4. Highlight D1:D9. Under the Formulas tab, select Create from Selection. Select Top Row. Click OK.

This gives the sample weights a name (Sample), so you can use them in the formulas you created for SaMean and StdDev. Notice B6 and B8 were filled in automatically.

5. Enter the remaining data in cells B5, B7, and B9 as shown below.

	A	B	C	D
1	Small Samples			Sample
2	Population Mean			60
3				56
4	Input Data			60
5	HoMean	60		55
6	SaMean	58.875		70
7	n	8		55
8	StdDev	5.083236		60
9	Alpha	0.1		55
10	Calculated Value			
11	df	7		
12				

After you are through entering the formulas and data, the values in the output cells have automatically changed to reflect the input data. Bold cells B12 and B15. You can now interpret the results.

Accept the H_0. This is a left-tailed test. The t value of $-.626$ (rounded) is greater than the left critical value of -1.415 (rounded). At the .10 significance there is not enough evidence to support the claim that the average salary of substitute teachers is less than $60 per day.

If you wish, save your file as **Ch8-prob2**. Close your file.

You can use this same template if the sample mean and standard deviation are given. Just key the values directly into B6 and B8.

	A	B	C	D
1	Small Samples			Sample
2	Population Mean			60
3				56
4	Input Data			60
5	HoMean	60		55
6	SaMean	58.875		70
7	n	8		55
8	StdDev	5.083236		60
9	Alpha	0.1		55
10	Calculated Value			
11	df	7		
12	t	**-0.62598**		
13	Test for Left-Tail			
14	LftCrt_tVal	-1.41492		
15	Conclusion	**Do Not Reject Ho**		
16	p-value	0.275594		
17	Test for Right-Tail			
18	RtCrt_tVal	1.414924		
19	Conclusion	Do Not Reject Ho		
20	p-value	0.724406		
21	Test for Two-Tail			
22	AbsCrt_tVal	1.894578		
23	Conclusion	Do Not Reject Ho		
24	p-value	0.551188		
25				

z Test for a Proportion

In the last exercise you created a worksheet used to conduct a test of hypothesis about a small population mean. You will again make some changes to a previously created worksheet. This worksheet will then be used to test a hypothesis about a single proportion. The formula for finding z is $z = \frac{\hat{p} - p}{\sqrt{pq/n}}$, where

$\hat{p}$ is the proportion in the sample possessing the trait. It will be referred to as *p_ratio*.

p is the hypothesized population proportion. It will be referred to as *p*.

q is *1-p*

n is the size of the sample. It will be referred to as *n*.

Example 8-4. Creating a worksheet for testing of hypothesis about a single proportion.

1. Retrieve the file **1sa-mean**.
2. Make A2 your active cell. Key **Single Proportion.**

3. Highlight A5:A8. Under the Home Tab click the Delete drop down arrow. Select Delete Sheet Rows.

This will delete the contents of the rows and the names of the cells that were created.

4. Highlight A5:A7. Under the Home Tab, click the Insert drop down arrow. Select Insert Sheet Rows.

This will give you blank rows to insert your input data.

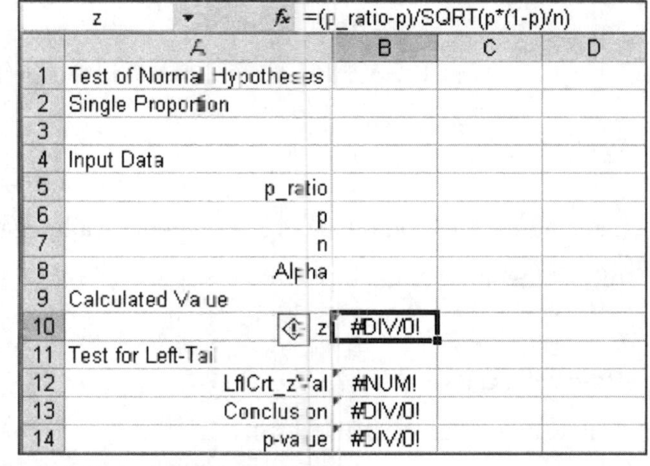

	A	B	C	D
1	Test of Normal Hypotheses			
2	Single Proportion			
3				
4	Input Data			
5	p_ratio			
6	p			
7	n			
8	Alpha			
9	Calculated Value			
10	z	#DIV/0!		
11	Test for Left-Tail			
12	LftCrt_zVal	#NUM!		
13	Conclusion	#DIV/0!		
14	p-value	#DIV/0!		

5. In A5:A7, key the input variables, as shown.

6. Highlight A5:A7. Under the Home Tab, select the Align Right icon.

7. Highlight A5:B7. Under the Formulas tab, select Create from Selection. Left Column. Click OK. Select Yes, for Replace Existing definition of "n"?

Excel® remembers *n* was used in an earlier worksheet. You are redefining *n*.

8. Make B10 your active cell. You will key the new formula for z. As you key, the previous formula for z will be replaced. Key **=(p_ratio-p)/SQRT(p*(1-p)/n)**

Save the workbook as **1-propor**.

Problem 8-3. An educator estimates that the dropout rate for seniors at high schools in Ohio is 15%. Last year, 38 seniors from a random sample of 200 Ohio seniors withdrew. At $\alpha = 0.05$, is there enough evidence to reject the educators claim?

1. Open the file **1-propor** if it is not already active.

2. In cell B5, key **=38/200**

By keying in the equation =38/200, you are computing the sample proportion.

3. In B6:B8, enter the input data in the appropriate cells as shown.

	A	B
1	Test of Normal Hypotheses:	
2	Single Proportion	
3		
4	Input Data	
5	p_ratio	0.19
6	p	0.15
7	n	200
8	Alpha	0.05
9		

The values in the output cells have automatically changed to reflect the input data.

Bold cells B10 and B21

You can now interpret the results.

	A	B	C
1	Test of Normal Hypotheses:		
2	Single Proportion		
3			
4	Input Data		
5	p_ratio	0.19	
6	p	0.15	
7	n	200	
8	Alpha	0.05	
9	Calculated Value		
10	z	**1.584236**	
11	Test for Left-Tail		
12	LftCrt_zVal	-1.64485	
13	Conclusion	Do Not Reject Ho	
14	p-value	0.94343	
15	Test for Right-Tail		
16	RtCrt_zVal	1.644853	
17	Conclusion	Do Not Reject Ho	
18	p-value	0.05657	
19	Test for Two-Tail		
20	AbsCrt_zVal	1.959961	
21	Conclusion	**Do Not Reject Ho**	
22	p-value	0.11314	
23			

This is a two-tailed test. Since the calculated value for z of 1.58 (rounded) is between the critical values of ± 1.96 (rounded), we do not reject the null hypothesis.

There is not enough evidence to reject the claim that the dropout rate is 15%.

If you wish, save your worksheet as **Ch8-prob3**.

Close your file.

Chi Square Test for a Variance

You will once again make changes to a previously created worksheet. This worksheet will be used to test a claim about a single variance using the chi-Square Test.

The formula for the chi-Square Test for a Single Variance is $x^2 = \frac{(n-1)s^2}{\sigma^2}$

Degrees of freedom equals n-1.

X^2 will be referred to as *Xsq*

n is the sample size and will be referred to as *n*

S^2 is the sample variance or the square of the standard deviation of the sample. It will be referred to as *Samp_Var*

σ^2 is the population variance or the square of the hypothesized standard deviation. It will be referred to as *Pop_Var*

Example 8-5. Creating a worksheet for the chi-Square Test for a Single Variance.

1. Open **1sa-mean**. Key **Test for Variance** in A1 and **Chi-Square** in A2.

2. Make A5 your active cell. Key **Samp_Var**

3. Highlight A5:B5, under the Formulas tab, select Create from Selection. Select Left Column. Click OK.

Check the name of cell B5 in the name box in the upper left corner. Sometimes the name does not change from the previous name. If the name in the name box still reads HoMean, do the following:

Highlight A5:B5, click Name Manager. Click on HoMean and select Delete. Click Close. Recheck the name of cell B5 in the name box. The name should read Samp_Var. Repeat this procedure anytime a name does not change to a new name, delete the old name.

4. Make A6 your active cell. Key **Pop_Var**

5. Highlight A6:B6. Repeat the instructions in step 3 for creating a name. Use the instructions under step 3 if the name does not change in the name box for cell B6.

6. In A8 key **df**.

7. In B8 key **=n-1**

8. Highlight A8:B8. Repeat the instructions in step 3 for creating a name.

9. In A11 key **Xsq**.

10. In B11 key **=((n-1)*Samp_Var)/Pop_Var**

11. Highlight A11:B11. Repeat the instructions in step 3 for creating a name.

12. Make A13 your active cell. Double click the left mouse button.

The cursor will appear in the cell. The cell can now be edited.

> Use the arrow keys to move the flashing vertical bar to the left of the letter z. Press the <delete> key once. Key the letters **Xsq**. (Remember to press the <Enter> key when you are finished with editing.)

Cell A13 should now read, LftCrt_XsqVal

13. Make B13 your active cell. Key **=CHIINV(1-Alpha,df)**

This replaces the old formula for LftCrt_zVal with the new formula for LftCrt_XsqVal.

14. Highlight A13:B13 Use the instructions in step 3 for creating a name.

15. Make B14 your active cell. Double click the left mouse button to edit the cell. Using your arrow keys to move the flashing vertical bar, delete the letter z both times and replace with the letters **Xsq**

Cell B14 should now read, =IF(Xsq<LftCrt_XsqVal, "Reject Ho", "Do Not Reject Ho")

16. Make B15 your active cell. Key **=1-CHIDIST(Xsq,df)**

This is the new p-value.

17. Make A17 your active cell. Double click the left mouse button. Use the arrow keys to move the flashing vertical bar to the left of the letter z. Press the <delete> key once. Key the letters **Xsq**.

Cell A17 should now read, RtCrt_XsqVal

18. Make B17 your active cell. Type **=CHIINV(Alpha, df)**

This replaces the old formula for RtCrt_zVal with the new formula for RtCrt_XsqVal

19. Highlight A17:B17 Use the instructions in step 3 for creating a name

20. Make B18 your active cell. Double click the left mouse button to edit the cell. Using your arrow keys to move the flashing vertical bar, delete the letter z both times and replace with the letters **Xsq**

Cell B18 should now read, =IF(Xsq>RtCrt_XsqVal, "Reject Ho", "Do Not Reject Ho")

21. . Make B19 your active cell. Key **=CHIDIST(Xsq,df)**

This gives the new p-value.

22. Make A21 your active cell. Key **Lft2tailVal**

23. Make B21 your active cell. Key **=CHIINV(1-(Alpha/2),df)**

24. Highlight A21:B21. Use the instructions in step 3 for creating a name. Check the name for cell B21. You may have to highlight A21:B21 and delete the name AbsCrt_zval, using the previous instructions.

25. Make A22 your active cell. Under the Home tab, click the Insert drop down arrow. Select Insert Sheet Rows.

26. In A22 key, **Rt2tailVal**.

27. Make B22 your active cell. Key **=CHIINV((Alpha/2),df)**

28. Highlight A22:B22 Use the instructions in step 3 for creating a name.

29. Make B23 your active cell. Key =IF(OR(Xsq<Lft2tailVal,Xsq>Rt2tailVal),"Reject Ho","Do Not Reject Ho").

If you wish you can double click the left mouse button on cell B23 and edit the cell to make the appropriate changes.

Cell B23 will now read. =IF(OR(Xsq<Lft2tailVal,Xsq>Rt2tailVal),"Reject Ho","Do Not Reject Ho")

30. Make cell B24 your active cell. Key **=2*(1-CHIDIST(Xsq,df))**

The cell contents of your worksheet will look as follows.

	A	B	C	D	E	F	G
1	Test for Variance						
2	Chi-Square						
3							
4	Input Data						
5	Samp_Var						
6	Pop_Var						
7	n						
8	df	=n-1					
9	Alpha						
10	Calculated Value						
11	Xsq	=((n-1)*Samp_Var)/Pop_Var					
12	Test for Left-Tail						
13	LftCrt_XsqVal	=CHIINV(1-Alpha,df)					
14	Conclusion	=IF(Xsq<LftCrt_XsqVal, "Reject Ho", "Do Not Reject Ho")					
15	p-value	=1-CHIDIST(Xsq, df)					
16	Test for Right-Tail						
17	RtCrt_XsqVal	=CHIINV(Alpha, df)					
18	Conclusion	=IF(Xsq>RtCrt_XsqVal,"Reject Ho", "Do Not Reject Ho")					
19	p-value	=CHIDIST(Xsq,df)					
20	Test for Two-Tail						
21	Lft2tailVal	=CHIINV(1-(Alpha/2),df)					
22	Rt2tailVal	=CHIINV((Alpha/2),df)					
23	Conclusion	=IF(OR(Xsq<Lft2tailVal,Xsq>Rt2tailVal),"Reject Ho", "Do Not Reject Ho")					
24	p-value	=2*(1-CHIDIST(Xsq,df))					
25							

Save your worksheet as **Chi-Square**.

Problem 8-4. An instructor wishes to see whether the variation in scores of the 23students in her class is less than the variance of the population. The variance of the class is 198. Is there enough evidence to support the claim that the variation of the students is less than the population variance ($\sigma^2 = 225$) at $\propto = .05$? Assume the scores are normally distributed.

1. Open the file **Chi-Square** if it is not already active.

2. In cell B5, key **198**; in cell B6, key **225**; in cell B7 key **23**; and in cell B9 key **.05**

The values in the output cells have automatically changed to reflect the input data.

Bold cells B11 and B14.

You can now interpret the results. Since the test value 19.36 is greater than the left critical value of 12.338 (rounded), the decision is to not reject the null hypothesis.

There is not enough evidence to support the claim that the variation in test scores of the instructor's students is less than the variation in scores of the population.

	A	B	C
1	Test for Variances		
2	Chi Square		
3			
4	Input Data		
5	Samp_Var	198	
6	Pop_Var	225	
7	n	23	
8	df	22	
9	Alpha	0.05	
10	Calculated Value		
11	Xsq	**19.36**	
12	Test for Left-Tail		
13	LftCrt_XsqVal	12.33801	
14	Conclusion	**Do Not Reject Ho**	
15	p-value	0.376994	
16	Test for Right-Tail		
17	RtCrt_XsqVal	33.92446	
18	Conclusion	Do Not Reject Ho	
19	p-value	0.623006	
20	Test for Two-Tail		
21	Lft2tailVal	10.98233	
22	Rt2tailVal	36.78068	
23	Conclusion	Do Not Reject Ho	
24	p-value	0.753989	
25			

If you wish, save your worksheet as **Ch8-prob4**.

Practice Exercises taken from textbook.

8-1. It has been reported that the average credit card debt for college seniors is $3262. The student senate at a large university feels that their seniors have a debt much less than this, so it conducts a study of 50 randomly selected seniors and finds that the average debt is $2995 with a sample standard deviation of $1100. With $\propto = 0.05$, is the student senate correct? (Textbook Exercise 8-3, Problem 2)

8-2. Full-time Ph.D. students receive an average salary of $12,837 according to the U.S. Department of Education. The dean of graduate studies at a large state university feels that Ph.D. students in his state earn more than this. He surveys 44 randomly selected students and finds their average salary is $14,445 with a standard deviation of $1500. With $\propto = 0.05$, is the dean correct? (Textbook Exercise 8-3, Problem 4)

8-3. The average salary for public school teachers for a specific year was reported to be $39,385. A random sample of 50 public school teachers in a particular state had a mean of $41,680 and a standard deviation of $5975. Is there sufficient evidence at the $\propto = 0.05$ level to conclude that the mean salary differs from $39,385? (Textbook Exercise 8-3, Problem 12)

8-4. A stockbroker thought that the average number of shares of stocks trades daily in the stock market was about 500 million. To test the claim, a researcher selected a random sample of 40 days and found the mean number of shares trades each day was 506 million shares. The standard deviation of the sample was 10.3. At $\propto = 0.05$, is there enough evidence to reject the broker's claim? (Textbook Chapter 8, Review Exercise # 2)

8-5. A state executive claims that the average number of acres in Western Pennsylvania State Parks is less than 2000 acres. A random sample of five parks is selected and the number of acres is shown. At $\alpha = 0.01$ is there enough evidence to support the claim? (Textbook Exercise 8-4, Problem 6)

959 1187 493 6249 541

8-6. The average undergraduate cost for tuition, fees, and room and board for 2-year institutions last year was $13,252. The following year, a random sample of 20 two-year institutions had a mean of $15,560 and a standard deviation of $3500. Is there sufficient evidence at the $\alpha = 0.01$ level to conclude that the mean cost has increased? Textbook Exercise 8-4, Problem 11)

8-7. Cushman and Wakefield reported that the average annual rent for office space in Tampa was $17.63 per square foot. A real estate agent selected a random sample of 15 rental properties (offices) and found the mean rent was $18.72 per square foot, and the standard deviation was $3.64. At $\alpha = 0.05$, test the claim that there is no difference in the rents. (Textbook Exercise 8-4, Problem 10)

8-8. Nationwide, the average salaries of actuaries who achieve the rank of Fellow is $150,000. An insurance executive wants to see how this compares with Fellows within his company. He checks the salaries of eight Fellows and finds the average salary to be $155,500 with a standard deviation of $15,000. Can he conclude that Fellows in his company make more than the national average, using $\alpha = 0.05$? (Textbook Chapter 8, Review Exercise # 4)

8-9. An item in *USA TODAY* reported that 63% of Americans owned an answering machine. A survey of 143 employees at a large school showed that 85 owned an answering machine. At $\propto = 0.05$, test the claim that the percentage is the same as stated in *USA TODAY*. (Textbook Exercise 8-5, Problem 9)

8-10. For a certain year a study reports that the percentage of college students using credit cards was 83%. A college dean of student services feels that this is too high for her university, so she randomly selects 50 students and finds that 40 of them use credit cards. Using $\propto = 0.04$, is she correct about her university? (Textbook Exercise 8-5, Problem 16)

8-11. The *Statistical Abstract* reported that 17% of adults attended a musical play in the past year. To test this claim, a researcher surveyed 90 people and found that 22 had attended a musical play in the past year. At $\propto = 0.05$, test the claim that this figure is correct. (Textbook Exercise 8-5, Problem 10)

8-12. According to the 2000 Census, 58.5% of women worked. A county commissioner feels that more women work in his county, so he conducts a survey of 1000 randomly selected women and finds that 622 work. Using $\propto = 0.05$ is he correct? (Textbook Chapter 8, Review Exercise # 8)

8-13. A researcher claims that the variance of the number of yearly forest fires in the United States is greater than 140. For a 13 year period, the variance of the number of forest fires in the United States is 146. Test the claim at $\propto = 0.01$. (Textbook Exercise 8-6, Problem 4)

8-14. Test the claim that the standard deviation of the number of aircraft stolen each year in the United States is less than 15 if a sample of 12 years had a standard deviation of 13.6. Use $\propto = 0.05$. (Textbook Exercise 8-6, Problem 5) Hint: Since the standard deviations are given, enter the square of each standard deviation for the variances.

8-15. A researcher claims that the standard deviation of the number of deaths annually from tornadoes in the United Stated is less than 35. If a sample of 11 randomly selected years had a standard deviation of 32, is the claim believable? Use $\propto = 0.05$. (Textbook Exercise 8-6, Problem 11) Hint: Since the standard deviations are given, enter the square of each standard deviation for the variances.

8-16. A film editor feels that the standard deviation for the number of minutes in a video is 3.4 minutes. A sample of 24 videos has a standard deviation of 4.2 minutes. At $\propto = 0.05$, is the sample standard deviation different from what the editor hypothesized? (Textbook Chapter 8, Review Exercise # 14) Hint: Since the standard deviations are given, enter the square of each standard deviation for the variances.

CHAPTER 9

TESTING THE DIFFERENCE BETWEEN TWO MEANS, TWO VARIANCES AND TWO PROPORTIONS

Chapter 9 is a continuation of hypothesis testing introduced in the previous chapter. As before, you will use Excel® to create templates to conduct the tests. After completing this chapter, you will be able to:

- Use the z-test to test the difference between two large sample means
- Test the difference between two variances, and two proportions
- Test the difference between two means for small independent, and small dependant samples

Steps in hypothesis testing:

These five steps were introduced in Chapter 8, and should provide continuity in completing hypothesis testing in this chapter.

1. **State the null and alternative hypothesis** using either formulas or words. The Null Hypothesis (H_o) is always the statement of no significant difference. The Alternative Hypothesis (H_1) is always the statement that there is a significant difference. When direction is stated it is a one-directional test (one-tailed). When direction is not stated it is a two-directional test (two-tailed).

2. **State the level of significance** or the probability that the null hypothesis is rejected when, in fact, it is true.

3. **State the statistical test** you will be using: the z test, t test, f test, chi square test, etc.

4. **Formulate a decision rule**. Using a picture or curve that estimates the distribution you are testing, show the critical value if you are performing a one-directional test or the upper and lower critical values if you are performing a two-directional test.

5. **Do it**. Show the formula you used and at least the major steps involved. State the results of the hypothesis test in terms of the question using complete sentences and examples.

Testing the Difference between Two Means: Large Samples

In the last chapter you created a worksheet used to conduct a test of hypothesis about a large sample mean from a population. You will make some changes to that worksheet. The next will be a test of hypothesis between sample means from two populations.

$$z = \frac{(\bar{x}_1 - \bar{x}_2) - (\mu_1 - \mu_2)}{\sqrt{\frac{s_1^2}{n_1} + \frac{s_2^2}{n_2}}}$$

The value of z, the critical value, is computed differently.

$\mu_1 - \mu_0$ is the expected difference.

$\overline{X}_1$ is the mean of the first sample. It will be referred to as X1_mean.

$\overline{X}_2$ is the mean of the second sample. It will be referred to as X2_mean.

s_1 is the standard deviation of the first sample. It will be referred to as s1_StdDev.

s_2 is the standard deviation of the second sample. It will be referred to as s2_StdDev.

n_1 is the first sample number. It will be referred to as n1_sample.

n_2 is the second sample number. It will be referred to as n2_sample.

Example 9-1. Creating a worksheet for testing of hypothesis between two population means.

1. Retrieve the file **1sa-mean**.

2. Make A2 your active cell. Key **Two Sample Means**

3. Highlight A5:A8. Under the Home Tab, click the Delete drop down arrow. Select Delete Sheet Rows.

This will delete the contents of the rows and the names of the cells that were created.

4. Highlight A5:A10. Under the Home Tab, click the Insert drop down arrow. Select Insert Sheet Rows.

This will give you blank rows to insert your input data.

5. In A5:A10, key the input variables, as shown.

	A
1	Test of Normal Hypotheses:
2	Two Sample Means
3	
4	Input Data
5	X1_mean
6	X2_mean
7	s1_StdDev
8	s2_StdDev
9	n1_sample
10	n2_sample
11	

6. Highlight A5:A10. From the Home tab, select the Align Right icon.

7. Highlight A5:B10. Under the Formulas tab, select Create from Selection. Select Left Column. Click on OK.

8. Make B13 your active cell. You will key a new formula for z. As you key, the formula created for z used with one sample mean will be replaced by the new formula. Use the shift key with the number 6 to key the ^ symbol. Key

=(X1_mean-X2_mean)/SQRT((s1_StdDev^2/n1_sample)+(s2_StdDev^2/n2_sample))

z | =(X1_mean-X2_mean)/SQRT((s1_StdDev^2/n1_sample)+(s2_StdDev^2/n2_sample))

	A	B	C	D	E	F	G	H	I
1	Test of Normal Hypotheses:								
2	Two Sample Means								
3									
4	Input Data								
5	X1_mean								
6	X2_mean								
7	s1_StdDev								
8	s2_StdDev								
9	n1_sample								
10	n2_sample								
11	Alpha								
12	Calculated Value								
13	z	#DIV/0!							

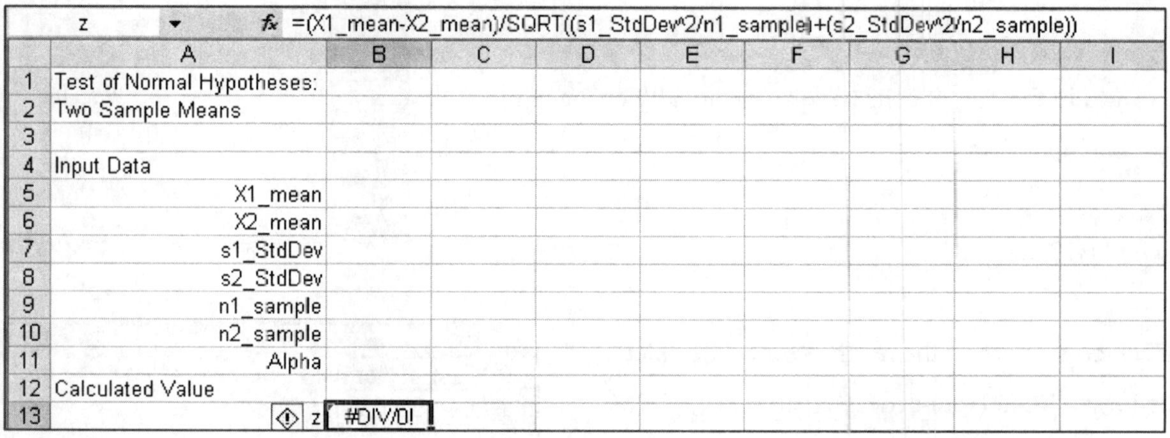

Save the worksheet as file name **2sa-mean**.

Problem 9-1. A survey found that the average hotel rate in New Orleans is $88.42 and the average room rate in Phoenix is $80.61. Assume that the data were obtained from two samples of 50 hotels each and that the standard deviations were $5.62 and $4.83 respectively. At $\propto = 0.05$, can it be concluded that there is a significant difference in the rates?

1. Open the file **2sa-mean** if it is not already active.

2. In B5:B11, enter the input data in the appropriate cells as shown.

	A	B
1	Test of Normal Hypotheses:	
2	Two Sample Means	
3		
4	Input Data	
5	X1_mean	88.42
6	X2_mean	80.61
7	s1_StdDev	5.62
8	s2_StdDev	4.83
9	n1_sample	50
10	n2_sample	50
11	Alpha	0.05
12		

The values in the output cells have automatically changed to reflect the input data.

Bold cell B13 and B24.

You can now interpret the results. Since the calculated value for z of 7.45 (rounded) is greater than +1.96 (rounded), we reject the null hypothesis. This is a two-tailed test. There is enough evidence to support the claim that the means are not equal and there is a significant difference in the hotel rates. The p-value shows that you have a .9.24E-14, chance of rejecting a true null hypothesis (a type one error).

	A	B	C
1	Test of Normal Hypotheses:		
2	Two Sample Means		
3			
4	Input Data		
5	X1_mean	88.42	
6	X2_mean	80.61	
7	s1_StdDev	5.62	
8	s2_StdDev	4.83	
9	n1_sample	50	
10	n2_sample	50	
11	Alpha	0.05	
12	Calculated Value		
13	z	**7.452419**	
14	Test for Left-Tail		
15	LftCrt_zVal	-1.64485	
16	Conclusion	Do Not Reject Ho	
17	p-value	1	
18	Test for Right-Tail		
19	RtCrt_zVal	1.644853	
20	Conclusion	Reject Ho	
21	p-value	4.62E-14	
22	Test for Two-Tail		
23	AbsCrt_zVal	1.959961	
24	Conclusion	**Reject Ho**	
25	p-value	9.24E-14	

If you wish, save your worksheet as **Ch9-prob1**. Close your file.

To use raw data, find the mean and standard deviation of the two sample means using the functions AVERAGE and STDEV. Input the data in the appropriate cells.

Testing the Difference between Two Variances

Excel® has a data analysis program for finding an f-test for two sample variances.

Problem 9-2. The CEO of an airport hypothesizes that the variance in the number of passengers for American airports is greater than the variance for the number of passengers for foreign airports. At $\propto = 0.10$ is there enough evidence to support the hypothesis? The data in millions of passengers per year is shown for selected airports. Assume the variable is normally distributed.

American airports		Foreign airports	
36.8	73.5	60.7	51.2
72.4	61.2	42.7	38.6
60.5	40.1		

	A	B
1	Am. AirP	Forn. AirP
2	36.8	60.7
3	72.4	42.7
4	60.5	51.2
5	73.5	38.6
6	61.2	
7	40.1	
8		

1. In a new worksheet, key the data as shown.

Note: The Analysis ToolPak option must be installed and selected before proceeding as noted Chapter 2 instructions.

2. Under the Data tab, select Data Analysis. Select F-Test Two Sample for Variances. Click OK.

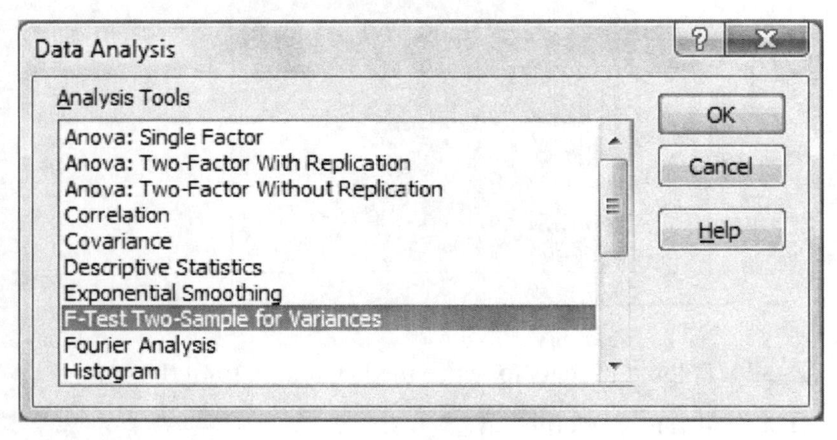

The F-Test Two-Sample for Variances dialog box is displayed.

3. Click inside the Variable <u>1</u> Range text box. Key **A1:A7**. Press the tab key.

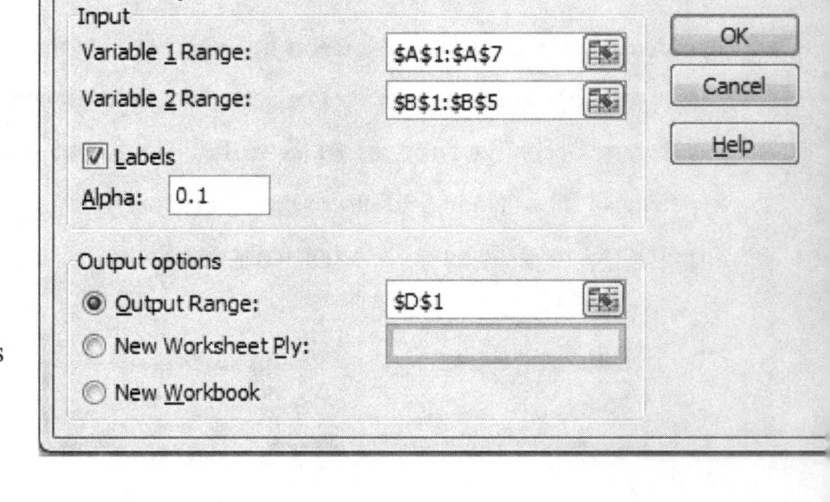

4. In the Variable 2 Range text box, key **B1:B5**. Press the tab key.

5. Select <u>L</u>abels, check box. Press the tab key.

6. In the <u>A</u>lpha text box key, **.10**

7. Select radio button for <u>O</u>utput Range. In the <u>O</u>utput Range text box, key **D1**. Click OK.

The output box is displayed but is difficult to read.

	A	B	C	D	E	F
1	Am. AirP	Forn. AirP		F-Test Two-Sample for Variances		
2	36.8	60.7				
3	72.4	42.7			Am. AirP	Forn. AirP
4	60.5	51.2		Mean	57.41667	48.3
5	73.5	38.6		Variance	246.3817	95.87333
6	61.2			Observatio	6	4
7	40.1			df	5	3
8				F	2.569866	
9				P(F<=f) on	0.233625	
10				F Critical c	5.309147	
11						

8. Double click between the D and E column headers to resize the D Column to fit the data. Your table in now easier to read and can be interpreted.

Bold cells E8 and E10. If you wish, save your file as **Ch9-prob2**. Close your file.

	A	B	C	D	E	F
1	Am. AirP	Forn. AirP		F-Test Two-Sample for Variances		
2	36.8	60.7				
3	72.4	42.7			Am. AirP	Forn. AirP
4	60.5	51.2		Mean	57.41667	48.3
5	73.5	38.6		Variance	246.3817	95.37333
6	61.2			Observations	6	4
7	40.1			df	5	3
8				F	**2.569866**	
9				P(F<=f) one-tail	0.233625	
10				F Critical one-tail	**5.309147**	
11						

The null hypothesis is not rejected at the .10 level because the computed value for F of 2.57 (rounded) is less than the critical value of 5.31 (rounded). There is not enough evidence to support the claim that the variance in the number of passengers for American airports is greater than the variance for the number of passengers for foreign airports.

Testing the Difference Between Two Means: Small Independent Samples

Excel® has a pre-prepared dialog box to use for t-tests with two sample means assuming unequal variances, using raw data. You will do the following problem using the Data Analysis dialog box:

Problem 9-3. Test the hypothesis that there is no difference between population means based on the following sample data. Assume the population variances are not equal. Use $\propto = 0.05$.

Set A	32	38	37	36	36	34	39	36	37	42
Set B	30	36	35	36	31	34	37	33	32	

1. In cells A1:B9 of a new worksheet, enter the data for Set A and Set B as shown.

	A	B
1	Set A	Set B
2	32	30
3	38	36
4	37	35
5	36	36
6	36	31
7	34	34
8	39	37
9	36	33
10	37	32
11	42	

2. Under the Data tab select Data Analysis. . Select t-Test: Two-Sample Assuming Unequal Variances in the Data Analysis dialog box. Click OK.

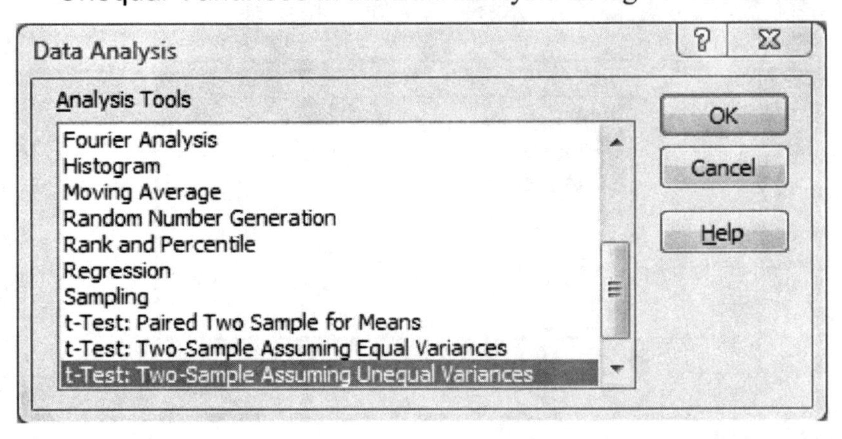

The dialog box for t-Test: Two-Sample Assuming Unequal Variances is displayed.

3. Your cursor should be on the Variable 1 Range text box: Key **A1:A11**. Press the tab key.

4. In the Variable 2 Range text box, key **B1:B10**. Press the tab key.

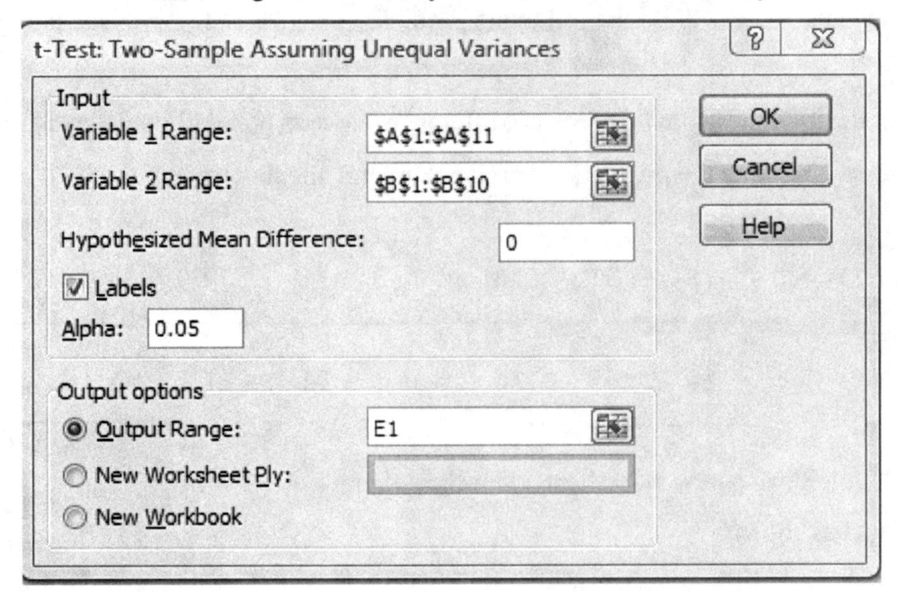

5. In the Hypothesized Mean Difference text box, key **0**. Press the tab key.

6. Select the Labels check box. Press the tab key.

7. In the Alpha text box, key **.05** if it is not already displayed.

8. Select the radio button for Output Range. In the text box, key **E1**. Click OK.

The output data is displayed, but is hard to read.

	A	B	C	D	E	F	G	H	I
1	Set A	Set B			t-Test: Two-Sample Assuming Unequal Variances				
2	32	30							
3	38	36				*Set A*	*Set B*		
4	37	35			Mean	36.7	33.77778		
5	36	36			Variance	7.344444	5.944444		
6	36	31			Observatio	10	9		
7	34	34			Hypothesi:	0			
8	39	37			df	17			
9	36	33			t Stat	2.474205			
10	37	32			P(T<=t) on	0.012095			
11	42				t Critical or	1.739606			
12					P(T<=t) tw	0.02419			
13					t Critical tv	2.109819			
14									

9. Make E7 your active cell. Under the Home tab select Format. Select AutoFit Column Width.

The output is now easier to read and can be interpreted. Bold cells F9 and F13.

	A	B	C	D	E	F	G
1	Set A	Set B			t-Test: Two-Sample Assuming Unequal Variances		
2	32	30					
3	38	36				*Set A*	*Set B*
4	37	35			Mean	36.7	33.77778
5	36	36			Variance	7.344444	5.944444
6	36	31			Observations	10	9
7	34	34			Hypothesized Mean Difference	0	
8	39	37			df	17	
9	36	33			t Stat	**2.474205**	
10	37	32			P(T<=t) one-tail	0.012095	
11	42				t Critical one-tail	1.739606	
12					P(T<=t) two-tail	0.02419	
13					t Critical two-tail	**2.109819**	
14							

Reject H_o. Since the computed value of t (2.474 rounded) is greater than the critical value of +2.110 (rounded) there is enough evidence to reject the claim that there is no difference between the population means. If you wish, save as **Ch9-prob3**. Do not close your file at this time if you plan to continue. You can use your worksheet for more than one problem. Make sure you use the same column for the output so the labels are in the same expanded column.

Testing the Difference Between Two Means: Small Dependent Samples (Paired Observations)
Excel® also has a dialog box to use with paired observations. You will do the following problem using the Data Analysis dialog box.

Problem 9-4. A physical education director claims by taking a special vitamin, a weight lifter can increase his strength. Eight athletes are selected and given a test of strength, using the standard bench press. After two weeks of regular training, supplemented with the vitamin, they are tested again. Test the effectiveness of the vitamin regimen at $\alpha = 0.05$. Each value in the data that follow represents the maximum number of pounds the athlete can bench press. Assume that the variable is approximately normally distributed.

Athlete	**1**	**2**	**3**	**4**	**5**	**6**	**7**	**8**
Before	210	230	182	205	262	253	219	216
After	219	236	179	204	270	250	222	216

1. On the same worksheet, beginning with row 16, enter the data in columns A, B, and C as shown below. If you had c osed your file, open **Ch9-prob3** and continue as

	A	B	C	D	E	F	G
1	Set A	Set B			t-Test: Two-Sample Assuming Unequal Variances		
2	32	30					
3	38	36				*Set A*	*Set B*
4	37	35			Mean	36.7	33.77778
5	36	36			Variance	7.344444	5.944444
6	36	31			Observations	10	9
7	34	34			Hypothesized Mean Difference	0	
8	39	37			df	17	
9	36	33			t Stat	**2.474205**	
10	37	32			P(T<=t) one-tail	0.012095	
11	42				t Critical one-tail	1.739606	
12					P(T<=t) two-tail	0.02419	
13					t Critical two-tail	**2.109819**	
14							
15							
16	Athlete	Before	After				
17	1	210	219				
18	2	230	236				
19	3	182	179				
20	4	205	204				
21	5	262	270				
22	6	253	250				
23	7	219	222				
24	8	216	216				
25							

shown below.

2. Select Data Analysis. Using the scroll bar on the right hand-side, select t-Test: Paired Two Samples for Means. Click OK.

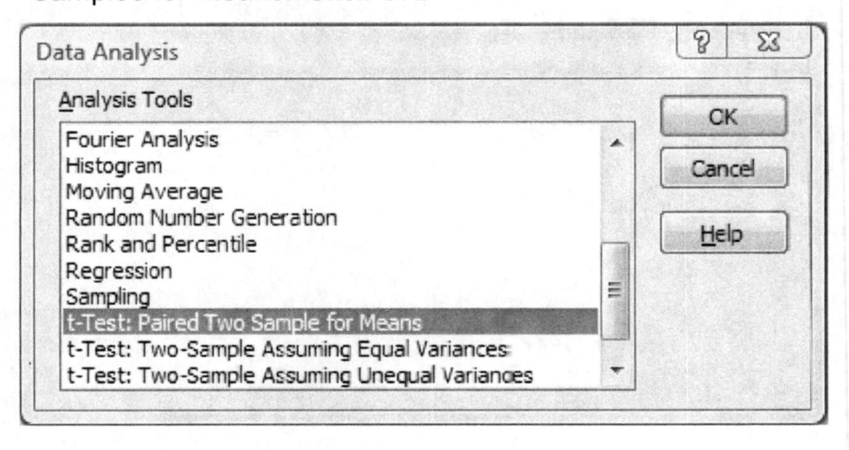

The dialog box for t-Test: Paired Two Sample for Means, is displayed.

3. With your cursor in the Variable 1 Range text box. Key **B16:B24**. Press the tab key.

4. In the Variable 2 Range text box, key **C16:C24**. Press the tab key.

5. In the Hypothesized Mean Difference text box, key **0**. Press the tab key.

6. Select the Labels check box. Press the tab key.

7. In the Alpha text box, key **.05** if it is not already displayed.

8. Select the radio button for Output Range. In the text box, key **E16**. Click OK.

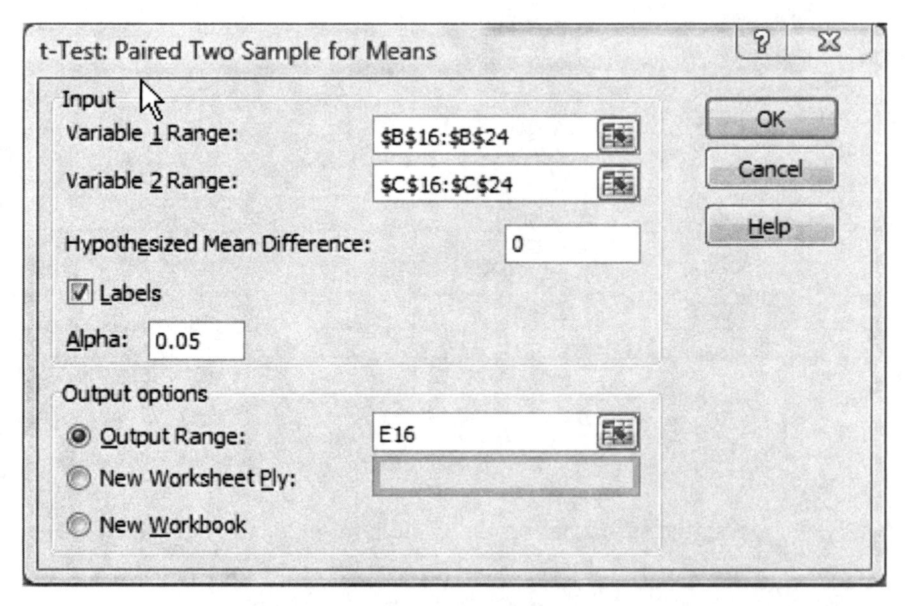

The output data is displayed. Bold cells F25 and F27.

	A	B	C	D	E	F	G
16	Athlete	Before	After		t-Test: Paired Two Sample for Means		
17	1	210	219				
18	2	230	236			*Before*	*After*
19	3	182	179		Mean	222.125	224.5
20	4	205	204		Variance	671.8393	778.8571
21	5	262	270		Observations	8	8
22	6	253	250		Pearson Correlation	0.986551	
23	7	219	222		Hypothesized Mean Difference	0	
24	8	216	216		df	7	
25					t Stat	**-1.38836**	
26					P(T<=t) one-tail	0.103802	
27					t Critical one-tail	**1.894578**	
28					P(T<=t) two-tail	0.207605	
29					t Critical two-tail	2.364623	
30							

Do not reject H_o. This is a left-tailed test. The computed t statistic (-1.388) is greater than the negative t Critical one-tail (-1.895). There is not enough evidence to support the claim that the vitamin increases the strength of weight lifters. (The critical t-value can be used for right or left tail. Use a negative value of the critical one-tail for left-tailed tests). If you wish, you may print both Problems 3 and 4.

	A	B	C	D	E	F	G
1	Set A	Set B			t-Test: Two-Sample Assuming Unequal Variances		
2	32	30					
3	38	36				*Set A*	*Set B*
4	37	35			Mean	36.7	33.77778
5	36	36			Variance	7.344444	5.944444
6	36	31			Observations	10	9
7	34	34			Hypothesized Mean Difference	0	
8	39	37			df	17	
9	36	33			t Stat	**2.474205**	
10	37	32			P(T<=t) one-tail	0.012095	
11	42				t Critical one-tail	1.739606	
12					P(T<=t) two-tail	0.02419	
13					t Critical two-tail	**2.109819**	
14							
15							
16	Athlete	Before	After		t-Test: Paired Two Sample for Means		
17	1	210	219				
18	2	230	236			*Before*	*After*
19	3	182	179		Mean	222.125	224.5
20	4	205	204		Variance	671.8393	778.8571
21	5	262	270		Observations	8	8
22	6	253	250		Pearson Correlation	0.986551	
23	7	219	222		Hypothesized Mean Difference	0	
24	8	216	216		df	7	
25					t Stat	**-1.38836**	
26					P(T<=t) one-tail	0.103802	
27					t Critical one-tail	**1.894578**	
28					P(T<=t) two-tail	0.207605	
29					t Critical two-tail	2.364623	
30							

If you wish, save as **Ch9-prob3&4**. Close your file.

A Test of Hypothesis about Two Population Proportions

The last large samples hypothesis with which you will be working is to conduct a test about two population proportions. You will make changes to a previously created worksheet one more time. The formula for finding z changes also.

$$z = \frac{(\hat{p}_1 - \hat{p}_2) - (p_1 - p_2)}{\sqrt{\overline{p}\overline{q}\left(\frac{1}{n_1} + \frac{1}{n_2}\right)}}$$

p_1-p_2 is the expected difference between two population proportions or 0.

$\overline{q}$ is 1- $\overline{p}$

$\hat{p}_1$ is the ratio of the first sample. It will be referred to as p1_ratio.

$\hat{p}_1$ is found by the formula $\hat{p}_1 = \frac{x_1}{n_1}$

X_1 is the number in the first sample that possess the trait. It will be referred to as X1_sample.

n_1 is the total number in the first sample. It will be referred to as n1_total.

$\hat{p}_2$ is the ratio of the second sample. It will be referred to as p2_ratio.

$\hat{p}_2$ is found by the formula $\hat{p}_2 = \frac{x_2}{n_2}$

X_2 is the number in the second sample that possess the trait. It will be referred to as X2_sample.

n_2 is the total number in the second sample. It will be referred to as n2_total.

$\overline{p}$ is the weighted estimate of the population proportion. It will be referred to as p_est.

$\overline{p}$ is found by the formula $\overline{p} = \frac{(X_1 + X_2)}{(n_1 + n_2)}$

Example 9-2. Creating a worksheet for testing of hypothesis about two population proportions.

1. Retrieve the file **1sa-mean**
2. Make A2 your active cell. Key **Two Population Proportions**
3. Highlight A5:A8. Under the Home Tab, click the Delete drop down arrow. Select Delete Sheet Rows.

This will delete the contents of the rows and the names of the cells that were created.

4. Highlight A5:A11. Under the Home Tab, click the Insert drop down arrow. Select Insert Sheet Rows.

This will give you blank rows to insert your input data.

5. In A5:A11, key the input variable names as shown.

6. Right Align A5:A11.

7. Create names for A5:B11.

8. Make B5 your active cell. Key **=X1_sample/n1_total**

9. Make B6 your active cell. Key =**X2_sample/n2_total**

	A
1	Test of Normal Hypotheses:
2	Two Population Proportions
3	
4	Input Data
5	p1_ratio
6	p2_ratio
7	X1_sample
8	X2_sample
9	n1_total
10	n2_total
11	p_est
12	Alpha
13	Calculated Value
14	

10. Make B11 your active cell. Key **= (X1_sample+X2_sample)/(n1_total+n2_total)**

11. Make B14 your active cell. As you key, the previous formula for z will be replaced. Key **=(p1_ratio-p2_ratio)/SQRT((p_est*(1-p_est)/n1_total)+(p_est*(1-p_est)/n2_total))**

z =(p1_ratio-p2_ratio)/SQRT((pc_est*(1-pc_est)/n1_total)+(pc_est*(1-pc_est)/n2_total))

	A	B	C	D	E	F	G	H	I
1	Test of Normal Hypotheses								
2	Two Population Proportions								
3									
4	Input Data								
5	p1_ratio	#DIV/0!							
6	p2_ratio	#DIV/0!							
7	X1_sample								
8	X2_sample								
9	n1_total								
10	n2_total								
11	pc_est	#DIV/0!							
12	Alpha								
13	Calculated Value								
14	z	#DIV/0!							

Save the worksheet as **2-propor**

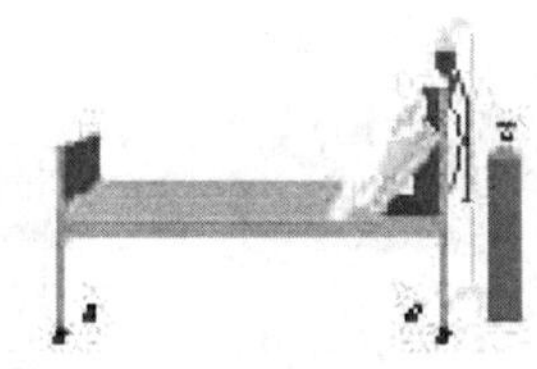

Problem 9-5. In the nursing home study mentioned in the textbook chapter-opening "Statistics Today," the researchers found that 12 out of 34 small nursing homes had a resident vaccination rate of less than 80%, while 17 out of 24 large nursing homes had a vaccination rate of less than 80%. At $\propto = 0.05$, test the claim that there is no difference in the proportions of the small and large nursing homes with a resident vaccination of less than 80%.

1. Open the file **2-propor** if it is not already active.

2. In B7:B10 and B12, enter the input data in the appropriate cells as shown.

	A	B
1	Test of Normal Hypotheses:	
2	Two Population Proportions	
3		
4	Input Data	
5	p1_ratio	#DIV/0!
6	p2_ratio	#DIV/0!
7	X1_sample	12
8	X2_sample	17
9	n1_total	34
10	n2_total	24
11	p_est	#DIV/0!
12	Alpha	0.05
13	Calculated Value	
14	z	#VALUE!
15		

The values in the output cells have automatically changed to reflect the input data. Bold cells B14 and B25.

	A	B	C
1	Test of Normal Hypotheses:		
2	Two Population Proportions		
3			
4	Input Data		
5	p1_ratic	0.352941	
6	p2_ratic	0.708333	
7	X1_sample	12	
8	X2_sample	17	
9	n1_total	34	
10	n2_total	24	
11	p_est	0.5	
12	Alpha	0.05	
13	Calculated Value		
14	z	**-2.66605**	
15	Test for Left-Tail		
16	LftCrt_zVal	-1.64485	
17	Conclusion	Reject Ho	
18	p-value	0.003837	
19	Test for Right-Tail		
20	RtCrt_zVal	1.644853	
21	Conclusion	Do Not Reject Ho	
22	p-value	0.996163	
23	Test for Two-Tail		
24	AbsCrt_zVal	1.959961	
25	Conclusion	**Reject Ho**	
26	p-value	0.007675	
27			

This is a two-tailed test.

You can now interpret the results. Reject the null hypothesis since –2.7 (rounded) is less than –1.96 (rounded). There is enough evidence to reject the claim that there is no difference in the proportions.

If you wish, save your worksheet as **Ch9-prob5**

Close your file.

Practice Exercises taken from textbook.

9-1. A study was conducted to see if there was a difference between spouses and significant others in coping skills when living with or caring for a person with multiple sclerosis. These skills were measured by questionnaire responses. The results of the two groups are given on one factor, ambivalence. At α = 0.10, is there a difference in the means of the two groups? (Textbook Exercise 9-2, Problem 6)

Spouses	Significant others
$\overline{X}_1 = 2.0$	$\overline{X}_2 = 1.7$
$s_1 = 0.6$	$s_2 = 0.7$
$n_1 = 120$	$n_2 = 34$

9-2. A survey of 1000 students nationwide showed a mean ACT score of 21.4. A survey of 500 Ohio scores showed a mean of 20.8. If the standard deviation in each case is 3, can we conclude that Ohio is below the national average? Use $\alpha = 0.05$. (Textbook Exercise 9-2, Problem 12)

9-3. Test the claim that the variance of heights of tall buildings in Denver is equal to the variance in heights of tall buildings in Detroit at $\alpha = 0.10$. The data are given in feet. (Textbook Exercise 9-3, Problem 17)

Denver			Detroit		
714	698	544	620	472	430
504	438	408	562	448	420
404			534	436	

9-4. The weights in ounces of a sample of running shoes for men and women are shown below. Test the claim that the variances are equal at $\alpha = 0.05$. (Textbook Exercise 9-3, Problem 19)

Men			Women		
11.9	10.4	12.6	10.6	10.2	8.8
12.3	11.1	14.7	9.6	9.5	9.5
9.2	10.8	12.9	10.1	11.2	9.3
11.2	11.7	13.3	9.4	10.3	9.5
13.8	12.8	14.5	9.8	10.3	11.0

9-5. A health care worker wishes to see if the average number of family day care homes per county is greater than the average number of day care centers per county. The number of centers for a selected sample of counties is shown. At $\alpha = 0.01$, can it be concluded that the average number of family day care homes is greater than the average number of day care centers? (Textbook Exercise 9-4, Problem 5)

Number of family day care homes			Number of day care centers		
25	57	34	5	28	37
42	21	44	16	16	48

9-6. A researcher wishes to test the claim that on average more juveniles than adults are classified as missing persons. Records for the last five years are shown. At $\alpha = 0.10$ is there enough evidence to support the claim? (Textbook Exercise 9-4, Problem 6)

Juveniles	65,513	65,934	64,213	61,954	59,167
Adults	31,364	34,478	36,937	35,946	38,209

9-7. A doctor is interested in determining whether a film about exercise will change 10 people's attitudes about exercise. The results of his questionnaire are shown below. A higher numerical value shows a more favorable attitude toward exercise. Is three enough evidence to support the claim at $\propto = .05$, that there was a change in attitude? Find the 95% confidence interval for the difference of the two means. (Textbook Exercise 9-5, Problem 4)

Before	12	11	14	9	8	6	8	5	4	7
After	13	12	10	9	8	8	7	6	5	5

9-8. In an effort to improve the vocabulary of 10 students, a teacher provides a weekly one-hour tutoring session for them. A pretest is given before the sessions and a posttest is given afterward. The results are shown in the table. At $\propto = 0.01$, can the teacher conclude that the tutoring sessions helped to improve the student's vocabularies? (Textbook Chapter 9, Review Exercise # 12)

Before	1	2	3	4	5	6	7	8	9	10
Pretest	83	76	92	64	82	68	70	71	72	63
Posttest	88	82	100	72	81	75	79	68	81	70

9-9. In Cleveland, a sample of 73 mail carriers showed that 10 had been bitten by an animal during one week. In Philadelphia, in a sample of 80 mail carriers, 16 had received animal bites. Is there a significant difference in the proportions? Use $\propto = 0.05$. Find the 95% confidence interval of the two proportions. (Textbook Exercise 9-6, Problem 6) (z value may differ because of rounding.)

9-10. St. Petersburg, Russia, has 207 foggy days out of 365 days while Stockholm, Sweden, has 166 foggy days out of 365. At $\propto = 0.02$, can it be concluded that the proportions of foggy days for the two cities are different? (Textbook Chapter 9, Review Exercise # 14)

CHAPTER 10
CORRELATION AND REGRESSION

The emphasis in this chapter is studying the relationship between two or more numbers. Regression analysis is a technique used to express the relationship between two variables that estimates the value of the dependent variable Y based on a selected value of the independent variable X. Correlation analysis is a group of statistical techniques used to measure the strength of the relationship (correlation) between two variables. This chapter expands the concept by allowing one to use more than one explanatory variable in a regression equation. Using more than one independent variable makes it possible to increase the explanatory power and the usefulness of regression and correlation analysis in making many business decisions. After completing this chapter, you will be able to:

- Use Excel® to draw a scatter plot.
- Use Excel® to calculate a regression line, coefficient of determination, and the correlation coefficient.
- Use Excel®'s Correl function to compute correlation coefficient.
- Make predictions using the regression equation.
- Use Excel® to find a multiple regression equation, predict a dependent variable based on two or more independent variables, and find the multiple coefficient of determination.

Scatter Plots and Regression Line

Problem 10-1. The data obtained in a study of age and systolic blood pressure of six randomly selected subjects is shown in the following table.

Subject	Age, x	Pressure, y
A	43	128
B	48	120
C	56	135
D	61	143
E	67	141
F	70	152

1-a. Construct a scatter plot for the data.

1-b. Find the equation of the regression line for the data.

1-c. Obtain the value of the correlation coefficient for the data.

1-d. Using the equation of the regression line, predict the blood pressure of a 50-year old person.

1-a. Creating a scatter plot

NOTE: When entering the data you must put the data for the independent variable **first**. The dependent variable, the variable being predicted or estimated, is shown on the vertical axis (Y-axis) and the independent variable, the predictor variable which provides the basis for estimation, is shown on the horizontal axis (X-axis).

	A	B	C
1	Subject	Age	Pressure
2	A	43	128
3	B	48	120
4	C	56	135
5	D	61	143
6	E	67	141
7	F	70	152
8			

1. On a new worksheet, enter the data for the problem as shown on the right. Notice that the data for Age is entered first since it is the independent variable.

2. Put your cursor on cell B2, click hold and drag to cell C7. Under the Insert tab, click on the Scatter icon and select the upper right chart.

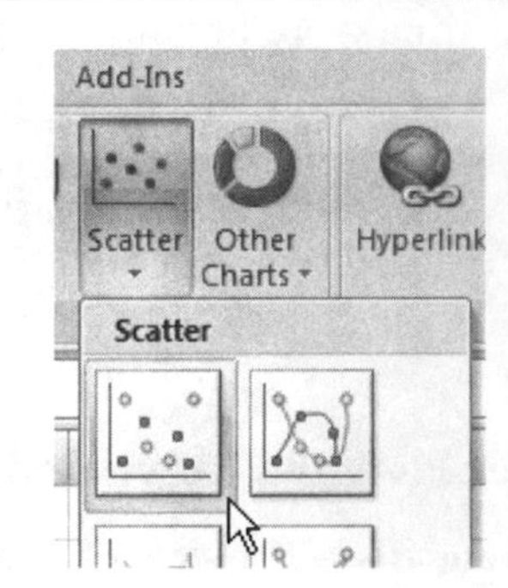

3. Under the Layout tab, click on the Chart Title icon and select Above Chart. Key **Correlation Between Age and Systolic Blood Pressure**. Press Enter.

4. Select the Layout tab and click the Axis Titles icon. Choose the following:
   a. Primary Horizontal Axis Title → Title Below Axis. Key **Age** to replace the default horizontal title. Press Enter.
   b. Primary Vertical Axis Title → Rotated Title. Key **Pressure** to replace the default vertical title. Press Enter.

5. Under the Layout tab, click the Legend icon. Select None.

1.a. The condensed scatter chart is formed.

	A	B	C
1	Subject	Age, x	Pressure, y
2	A	43	128
3	B	48	120
4	C	56	135
5	D	61	143
6	E	67	141
7	F	70	152

Correlation Between Age and Systolic Blood Pressure

Pressure: 0, 50, 100, 150, 200

Age: 0, 20, 40, 60, 80

1-b & c. Obtaining the equation of the regression line and the value of correlation coefficient

You will now add a regression line or a *trendline*. The regression equation ($y' = a + bx$) is the mathematical equation that defines the relationship between two variables that have a linear relationship.

a is the estimated value of the dependent variable Y where the regression line crosses the Y-axis when X is zero.

b is the slope of the line, or the average change in y' for each change of one unit (either increase or decrease) in the independent variable X.

1. Make sure the frame shows around the chart box. With your **right** mouse arrow, click on one of the data points. Open dots will appear around each data point.

2. From the pull down list, select Add Trendline

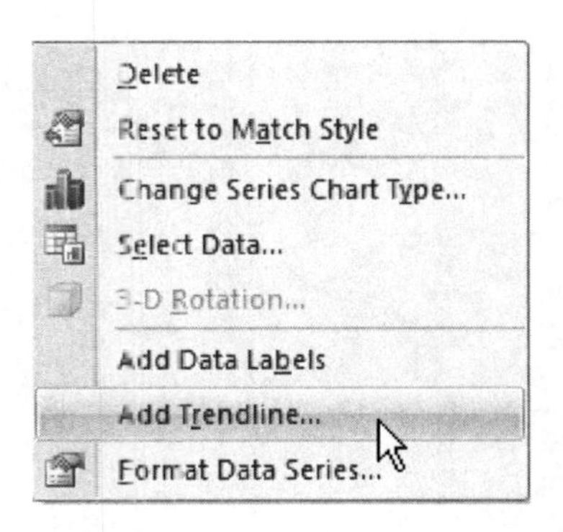

The Add Trendline dialog box appears.

3. Under the Type tab, Linear should be selected for Trend/Regression type. Click on the Options tab.

The Add Trendline Options dialog box appears.

4. Linear should be selected under Trend/Regression Type and Automatic should be selected under Trendline name. Select the check boxes for Display Equation on chart and Display R-squared value on chart. Click Close.

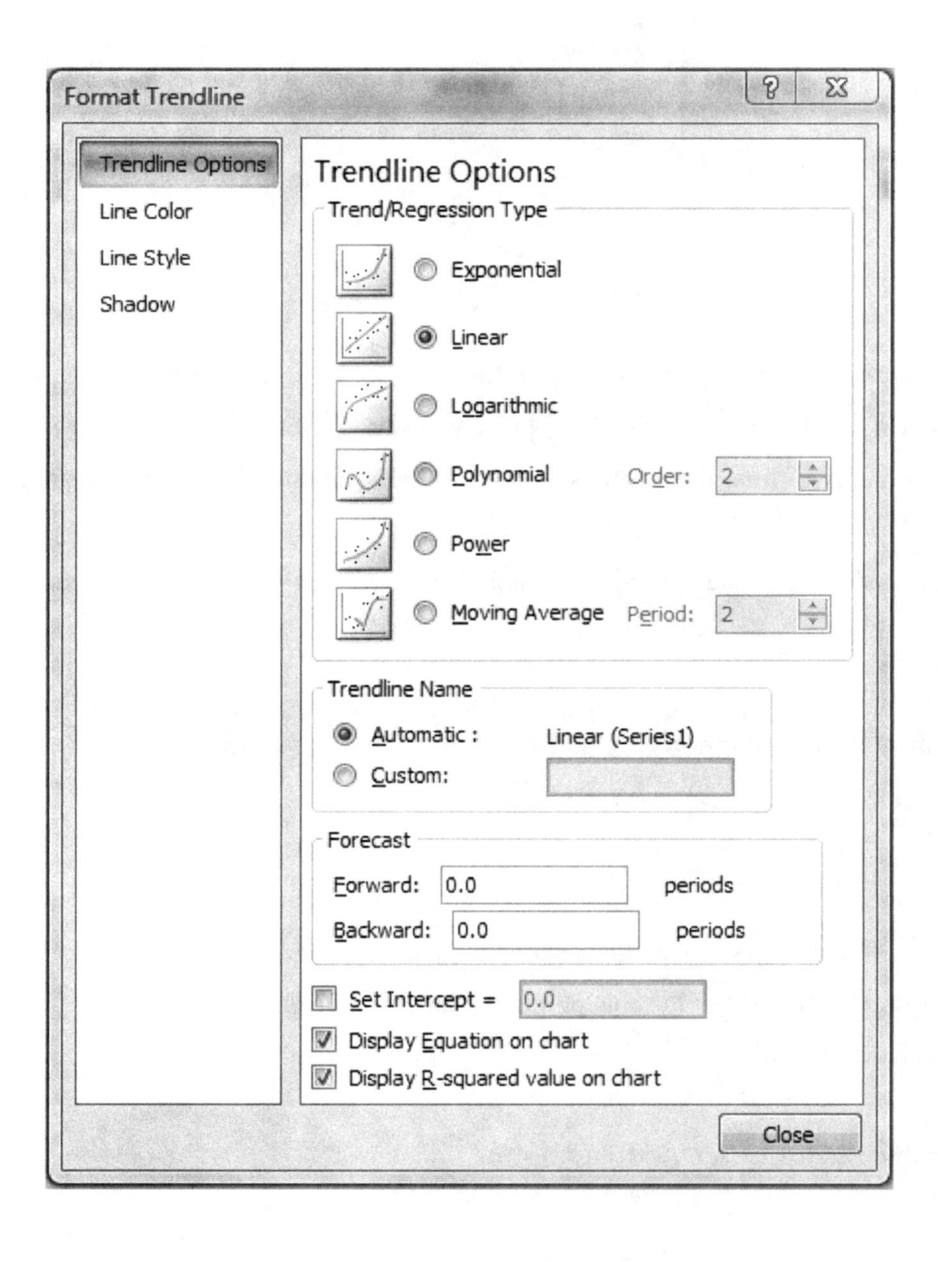

The regression equation and R^2 are shown, but are difficult to read. The coefficient of determination, R^2, is the total variation in the dependent variable Y that is explained, or accounted, for by the variation in the independent variable X. To find the correlation coefficient, take the square root of the coefficient of determination.

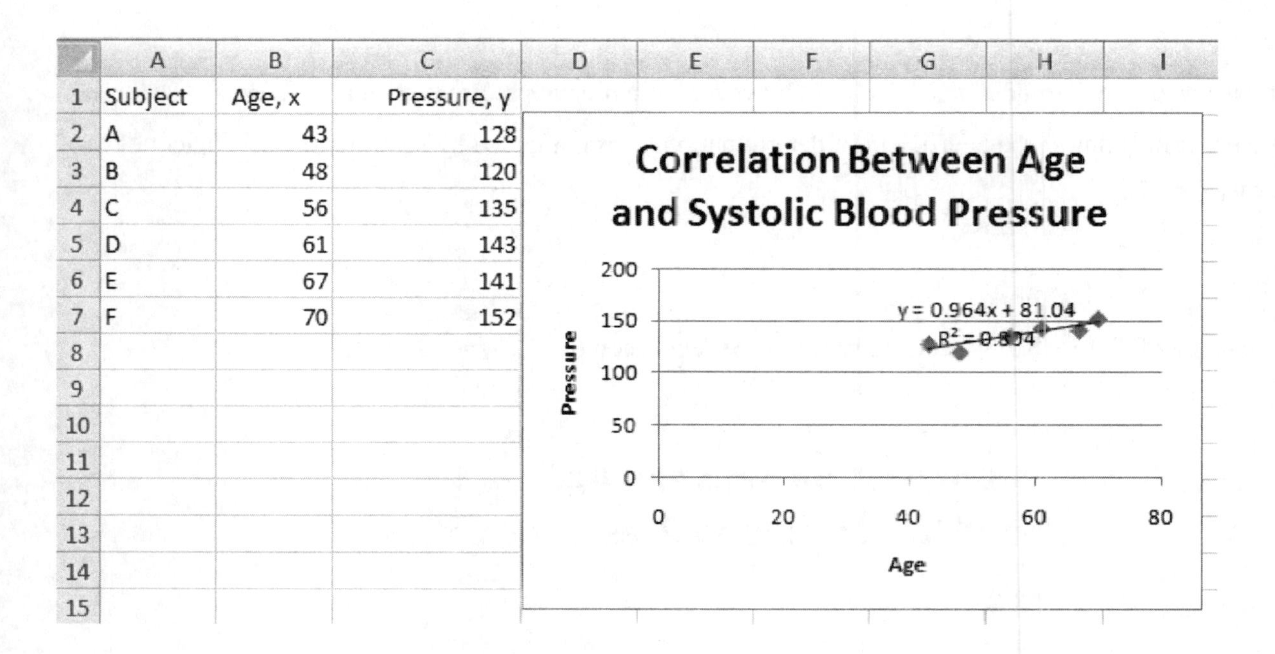

	A	B	C
1	Subject	Age, x	Pressure, y
2	A	43	128
3	B	48	120
4	C	56	135
5	D	61	143
6	E	67	141
7	F	70	152

5. Click anywhere on the equation. A box appears around the outside Place your arrow on any line. Your mouse arrow will remain as an arrow. Click, hold, and drag the box to the left bottom corner of the graph.

The equation y=0.964x +81.04 and $R^2 = 0.804$ are now easier to read.

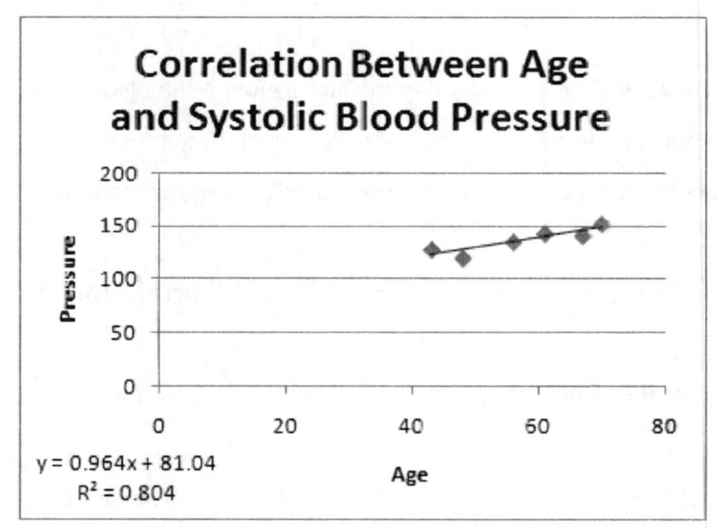

The correlation coefficient may assume any value on a scale of -1 to +1, inclusive. It describes the strength of the relationship between two sets of interval-scaled or ratio-scaled variables. A correlation coefficient of -1 or +1 indicates a perfect correlation. A coefficient of correlation close to 0 shows a weak relationship.

Terms such as weak, moderate, and strong, however, do not have precise meaning. A measure that has a more exact meaning is the coefficient of determination. It is computed by squaring the coefficient of correlation.

1-b(cont). y=0.964x +81.04 is the equation of the regression line.

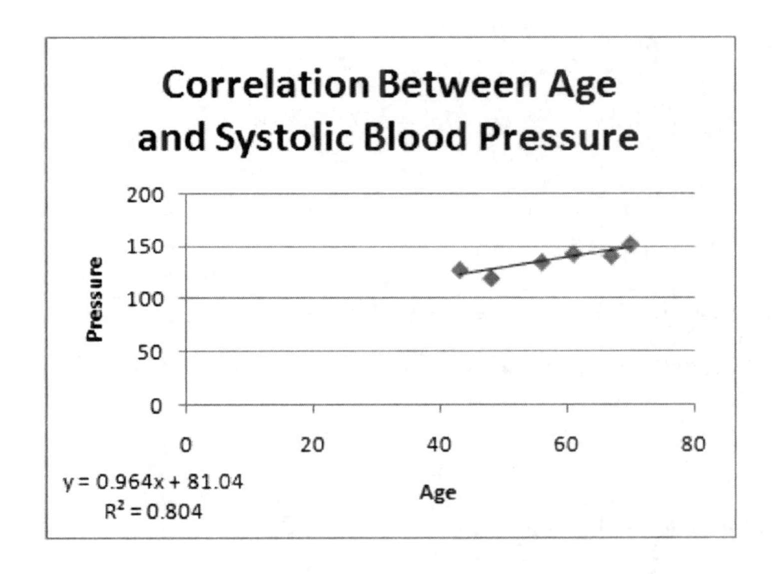

1-c(cont). Since you requested Excel® to display the R^2 value on the chart, you can read from the chart that the coefficient of determination is .804, which is the correlation coefficient squared. To find the correlation coefficient, take the square root of the coefficient of determination.

6. On the same worksheet containing your chart, in A9, key **Correlation Coeffic** =

7. In cell C9 key, **=SQRT(.804)**. The correlation coefficient is .89666.

Since the correlation coefficient is close to –1 or +1 we will assume a strong linear relationship.

1-d. Using the equation of the regression line to make predictions

You will use the regression equation to predict the blood pressure for a person who is 50 years old (x). The equation is $y' = a + bx$, or $y = bx + a$, as shown on the chart.

Since you requested Excel® to display the equation on the chart, you can use the given equation for your prediction. The equation on the chart uses "y" instead of "y'" and has "a" and "bx" in reverse order, but otherwise it is the same equation. 81.048 is the value of "a" and .9644 is the value of "b".

1. On the same worksheet containing your chart, in cell A11, key **Age**. In cell B11, key **Predicted pressure**, in cell A12, key **50**

2. In B12, key **=0.964*A12+81.048**

3. Decrease the decimal to round to a whole number.

This computes the predicted pressure for a person age 50, 129 (rounded). Since you created a formula, you can use cell B12 to predict the blood pressure for any other age. Just key the age in cell A12, and the formula will automatically compute the predicted blood pressure.

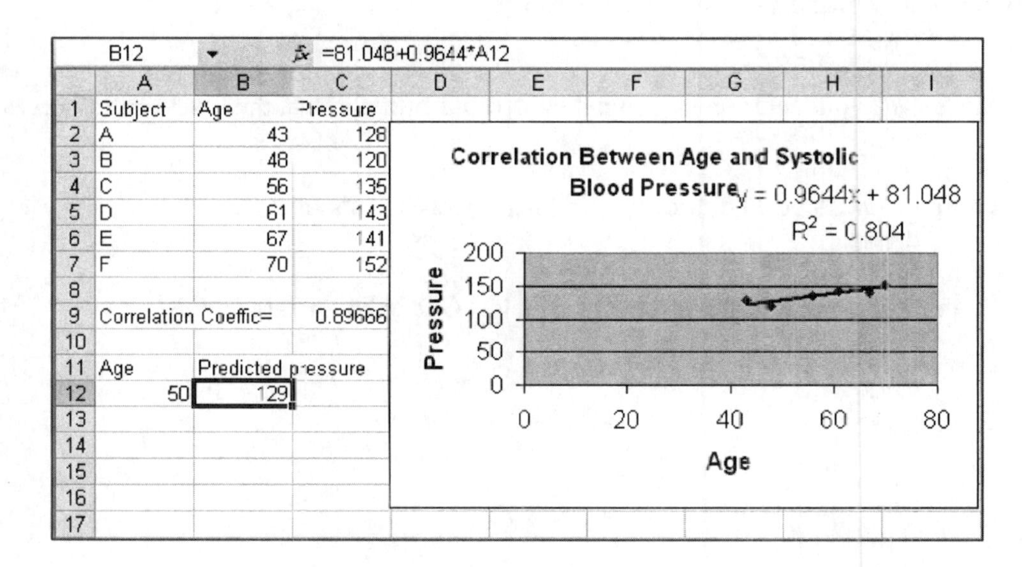

If you wish, save your file as **Ch10-prob1**. Close your file.

Using Excel®'s CORREL function to compute correlation coefficient

Problem 10-2. The data obtained in a study on the number of hours nine people exercise each week and the amount of milk (in ounces) each person consumes per week is shown below. Find the correlation coefficient.

Subject	Hours,x	Amount,y
A	3	48
B	0	8
C	2	32
D	5	64
E	8	10
F	5	32
G	10	56
H	2	72
I	1	48

	A	B	C
1	Subject	Hours	Amount
2	A	3	48
3	B	0	8
4	C	2	32
5	D	5	64
6	E	8	10
7	F	5	32
8	G	10	56
9	H	2	72
10	I	1	48
11			
12	Correlation Coeffi=		
13			

1. On a new worksheet enter the data as shown.

2. Make C12 your active cell. From the Formula bar, click on the Insert Function icon. 

3. From the Or select a category scroll bar select Statistical.

4. From the Select a function scroll bar, select CORREL. Click on OK.

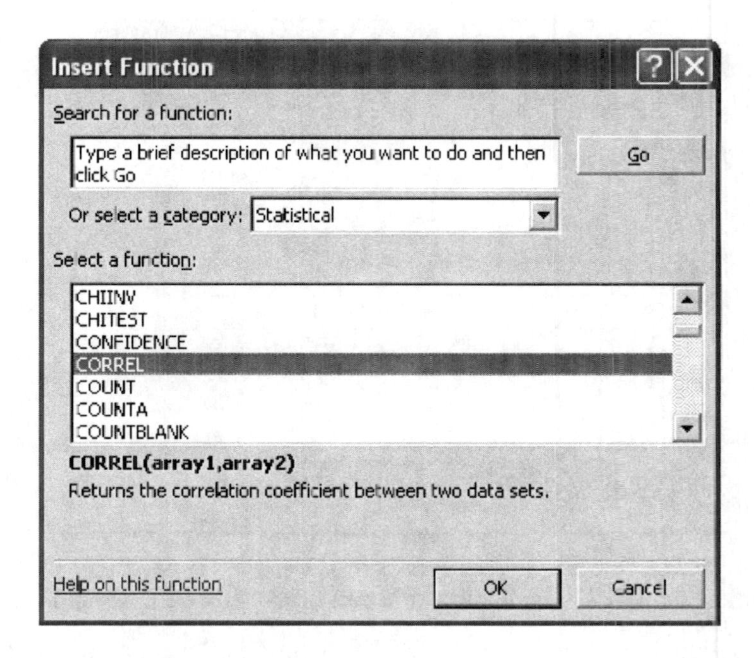

5. In the Array 1 text box, click on the icon at the far right. Put your cursor on cell B2, click, hold, and drag to cell B10. There will be a running box around cells B2:B10. Press your <Enter> key.

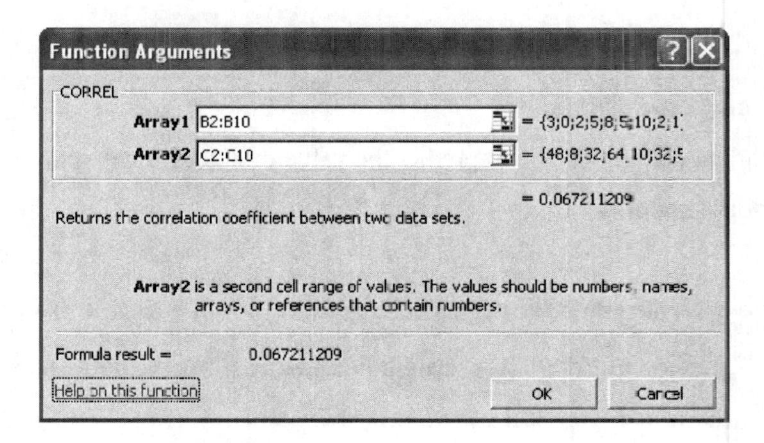

6. In the Array 2 text box, click on the icon at the far right. Put your cursor on cell C2, click, hold, and drag to cell C10. There will be a running box around cells C2:C10. Press your <Enter> key. Click OK.

C12		f_x =CORREL(B2:B10,C2:C10)		
	A	B	C	D	E

	A	B	C	D	E
1	Subject	Hours	Amount		
2	A	3	48		
3	B	0	8		
4	C	2	32		
5	D	5	64		
6	E	8	10		
7	F	5	32		
8	G	10	56		
9	H	2	72		
10	I	1	48		
11					
12	Correlation Coeff=		0.067211		
13					

The correlation coefficient of .067211 is displayed in cell C12. Since the value is close to "0" we will assume a very weak relationship. The Excel® CORREL function allows you to compute the correlation coefficient without doing a scatter diagram and trendline.

If you wish save your file as **Ch10-prob2**

Multiple Regression and Correlation

The multiple regression model allows one to predict the value of a dependent variable by incorporating two or more independent variables.

The textbook states the estimated multiple regression model as $y' = a + b_1x_1 + b_2x_2 + \ldots b_kx_k$. Excel® uses the "least squares" method to calculate a straight line that best fits the data and returns an array that describes the line. The multiple regression equation for the line is: $y = m_1x_1 + m_2x_2 + \ldots m_nx_n + b$.

You will use the LINEST function of Excel®. LINEST gives you the means to predict the dependent variable and the coefficient of determination.

The LINEST function is written, **=LINEST(known_y's,known_x's,const,stats)** , where:

known_y's is the range of y-values you already know

known_x's is the range of the known variables

const is a logical value specifying whether or not to calculate the constant normally. For these problems it will always be TRUE.

stats is a logical value specifying whether to return additional regression statistics such as the coefficient of determination. Since you want the additional statistics, it will always be TRUE.

The *array* that LINEST *returns* is $\{m_n, m_{n-1},...m_1, b\}$. So the coefficients that correspond to each x-value in the array are the **reverse** of the coefficients in the multiple regression equation for the line. When entering the data into Excel®, you must enter the data for the independent variables **first**, then your known y values.

Problem 10-3.

The nursing instructor wishes to see whether a student's grade point average and age are related to the student's score on the state board nursing examination. She selects five students and obtains the following data.

Student	GPA x1	Age x2	St. Board score y
A	3.2	22	550
B	2.7	27	570
C	2.5	24	525
D	3.4	28	670
E	2.2	23	490

Using a multiple regression equation, what is the predicted state board score for a student with a GPA of 3.0 and who is 25 years old.

You will use the LINEST function to calculate coefficients needed in the multiple regression to predict the state board score. Notice on the following worksheet, the independent variables of x2 and x1 are entered first, then the known y values. In the multiple regression line, Excel® reverses the order of the coefficients from the way they were listed in the worksheet. If you enter the independent variables in reverse order they will be correct in the multiple regression equation.

1. On a new worksheet, enter the data as shown.

	A	B	C	D
1	Multiple regression			
2				
3		Age, x2	GPA, x1	score, y
4	A	22	3.2	550
5	B	27	2.7	570
6	C	24	2.5	525
7	D	28	3.4	670
8	E	23	2.2	490
9				
10		GPA, x1	Age, x2	Constant
11				
12				
13	coef of det			
14				

Notice that where the answers will be displayed, you place the coefficients in reverse order of the way they were entered on the worksheet. The LINEST function always displays the constant value last. Be sure when you key the LINEST function you list the cell contents for the known **Y values first**.

2. Highlight B11:D13. With the range still highlighted, key
   =LINEST(D4:D8,B4:C8,TRUE,TRUE)

DO NOT PRESS THE <ENTER> KEY YET!

3. After you have finished keying, hold down the <Shift> key and the <Ctrl> key together and at the same time press the <Enter> key. The formula in the formula bar at the top of the worksheet should be inside curly brackets, {}.

B13		fx {=LINEST(D4:D8,B4:C8,TRUE,TRUE)}				
	A	B	C	D	E	F
1	Multiple regression					
2						
3		Age, x2	GPA, x1	score,y		
4	A	22	3.2	550		
5	B	27	2.7	570		
6	C	24	2.5	525		
7	D	28	3.4	670		
8	E	23	2.2	490		
9						
10		GPA, x1	Age, x2	Constant		
11		87.64015	14.53297	-44.8102		
12		15.23719	2.913738	69.24687		
13	coef of det	0.978691	14.00909	#N/A		
14						

Notice the coefficients for the independent variables are reversed. The coefficient for x1 (87.640) is displayed in cell **B11**, the coefficient for x2 (14.533) is displayed in cell **C11**, and the constant value (-44.810) is displayed in cell **D11**. The coefficient of determination is displayed in cell **B13**.

Always highlight 3 rows before you key your LINEST function. In the Excel® output, the coefficient of determination will be the first cell in the third row.

Using Row 11, you can construct a multiple regression equation to predict the state board score (y):
Y = -44.8102 + 87.64015 (GPA) + 14.53297 (Age)

4. In A15 to A17, key respectively, **GPA(x1)=, Age(x2)=, Predicted score(y)=**

5. In C15, key **3.0**, in C16, key **25**, in C17, key **=D11+ C15*B11 + C16*C11**. Press Enter.

C17		f_x =D11+C15*B11+C16*C11			
	A	B	C	D	E
1	Multiple regression				
2					
3		Age, x2	GPA, x1	score,y	
4	A	22	3.2	550	
5	B	27	2.7	570	
6	C	24	2.5	525	
7	D	28	3.4	670	
8	E	23	2.2	490	
9					
10		GPA, x1	Age, x2	Constant	
11		87.64015	14.53297	-44.8102	
12		15.23719	2.913738	69.24687	
13	coef of det	0.978691	14.00909	#N/A	
14					
15	GPA(x1)=		3		
16	Age(x2)=		25		
17	Predicted score(y)=		581.4346		
18					

Decrease the decimal of C17 to a whole number.

This gives you the estimated state board score of 581.4346 or 581 (rounded). If you want to experiment to see what the predicted state board score would be for different values of GPA and age, you can key in other values for cells C15 and C16.

If you wish, save your file as **Ch10-prob3**

Practice Exercises taken from the textbook.

10-1. An environmentalist wants to determine the relationships between the numbers (in thousands) of forest fires over the year and the number (in hundred thousands) of acres burned. The data for 8 recent years are shown. Draw a scatter diagram, find the equation of the regression line, the correlation coefficient, and predict the number of acres burned (in hundred thousands) when there are 60 forest fires. (Textbook Exercise 10-3, problem 14 and Exercise 10-4, Problem 14)

No. of fires, x,	72	69	58	47	84	62	57	45
No of acres burned, y,	62	42	19	26	51	15	30	15

10-2. An emergency service wished to see whether a relationship exists between the outside temperature and the number of emergency calls it receives for a 7-hour period. The data are shown. Draw a scatter diagram, find the equation of the regression line, the correlation coefficient, and predict the number of calls when the outside temperature is 80 degrees. (Textbook Exercise 10-3, Problem 20 and Exercise 10-4, Problem 20)

Temperature, x,	68	74	82	88	93	99	101
No. of calls, y	7	4	8	10	11	9	13

10-3. A store manager wishes to find out whether there is a relationship between the age of her employees and the number of sick days they take each year. The data for the sample follow. Draw a scatter diagram, find the equation of the regression line, the correlation coefficient, and predict the number of sick days for an employee age 47 years. (Textbook Exercise 10-3, Problem 16 and Exercise 10-4, Problem 16)

Age, x	18	26	39	48	53	58
Days, y	16	12	9	5	6	2

10-4. A study was conducted with vegetarians to see whether the number of grams of protein each ate per day was related to diastolic blood pressure. The data are given here. Draw a scatter diagram, find the equation of the regression line, the correlation coefficient, and predict the diastolic pressure of a vegetarian who consumes 8 grams of protein per day. (Textbook Chapter 10, Review Exercise # 6)

Grams, x	4	6.5	5	5.5	8	10	9	8.2	10.5
Pressure, y	73	79	83	82	84	92	88	86	95

CHAPTER 11
OTHER CHI-SQUARE TESTS

Chapters 8 through 10 dealt with data that were at least interval-scale, such as weights, incomes, and ages. We conducted a number of tests of hypothesis about a population mean and about two or more population means. For these tests it was assumed that the population was normally distributed. This chapter deals with the chi-square distribution, hypothesis tests where the data does not need to be interval-scale, but could be nominal- or ordinal-scale, and where no assumptions are made about the shape of the population distribution. This type of hypothesis test is called nonparametric (further discussion of nonparametric testing will continue in Chapter 13.) In this type of test, the number of degrees of freedom is determined by k-1, where k is the number of categories. Therefore, the shape of the chi-square distribution does not depend on the size of the sample, but rather as the number of degrees of freedom increases, the distribution begins to approximate the normal curve. After completing this chapter, you will be able to:

- Use Excel®'s IF function to conduct a goodness-of-fit test of hypothesis involving the difference between a set of observed frequencies and a corresponding set of expected frequencies.
- Use Excel® to conduct a contingency table analysis, a test of hypothesis to determine whether two criteria of classification are related.

Equal Expected Frequencies

In this chapter you will be working with the chi-square distribution. There is no worksheet that can be created that will fit all problems of a particular type. What follows are examples of different situations. You will need to modify the worksheets when doing other problems.

The formula, referred to as chi_sq, for the test statistic of chi-square is $X^2 = \Sigma\left[\frac{(O - E)^2}{E}\right]$ where,

O is an observed frequency in a particular situation. It will be referred to as fo.
E is an expected frequency in a particular situation. It will be referred to as fe.

You will also need the degrees of freedom for computing the critical value. Degrees of freedom will be referred to as df, the critical value will be referred to as Crit_Val. The formula for degrees of freedom is df = k-1, where k is the number of categories or *cells*.

You will continue to use the five steps in hypothesis testing:

1. **State the null and alternative hypothesis** using either formulas or words. The Null Hypothesis (H_o) is always the statement of no significant difference. The Alternative Hypothesis (H_1) is always the statement that there is a significant difference. When direction is stated it is a one-directional test (one-tailed). When direction is not stated it is a two-directional test (two-tailed).

2. **State the level of significance** or the probability that the null hypothesis is rejected when, in fact, it is true.

3. **State the statistical test** you will be using: the z test, t test, f test, chi-square test, etc.

4. **Formulate a decision rule**. Using a picture or curve that estimates the distribution you are testing, show the critical value if you are performing a one-directional test or the upper and lower critical values if you are performing a two-directional test.

5. **Do it**. Show the formula you used and at least the major steps involved. State the results of the hypothesis test in terms of the question using complete sentences and examples.

Goodness of Fit, Example for 5 categories

Problem 11-1. A market analyst wishes to see whether consumers have any preference among five flavors of a new fruit soda. A sample of 100 people provided these data. Is there evidence to reject the claim that there is no preference in the selection of fruit soda flavors? Let $\alpha = 0.05$.

Soda flavor	Cherry	Strawberry	Orange	Lime	Grape
Soda consumption	32	28	16	14	10

1. On a new worksheet, enter the data as shown.

	A	B	C	D
1	Chi-Square	Equal Expected Frequencies		
2				
3		fo	fe	ChiSqDst
4	Cherry	32		
5	Strawberry	28		
6	Orange	16		
7	Lime	14		
8	Grape	10		
9				
10	k			
11	Chi_Sq			
12	df			
13	Alpha			
14	Crit_Val			
15	Conclusion			
16				

2. Highlight A4:A14. Under the Home tab, click the Align Right icon.

3. Highlight B3:D8. Under the Formulas tab, click Create from Selection. Select Top Row. Click OK.

4. Highlight A10:B14. Under the Formulas tab, click Create from Selection. Select Left Column. Click OK.

5. In B10, key **=COUNT(fo)**

This will count the number of categories and give you the value of k.

6. In B11, key **=SUM(ChiSqDst)**

This will sum all the cell contents that you calculate.

7. In B12, key **=k-1**

8. In B13, key **.05**

9. In B14, key **=CHIINV(Alpha, df)**

This computes the critical value.

10. In B15, key **=IF(Chi_Sq>Crit_Val, "Reject Ho", "Do Not Reject Ho")**

As you are keying, your cell contents for A10:B15 will look as shown.

Ignore the results in cells B14 and 15 for now.

	A	B	C	D	E	F
10	k	=COUNT(fo)				
11	Chi_Sq	=SUM(ChiSqDst)				
12	df	=k-1				
13	Alpha	0.05				
14	Crit_Val	=CHIINV(Alpha, df)				
15	Conclusior	=IF(Chi_Sq>Crit_Val, "Reject Ho", "Do Not Reject Ho")				
16						

11. In C4, key **=AVERAGE(fo)**

This is the expected frequency (f_e). This value will later be copied to cells C5:C8.

12. In D4, key **=(fo-fe)^2/fe**

This is the square of each difference divided by the expected frequency. This formula will later be copied to cells D5:D8.

The cell contents of C4 and D4 will look as shown.

	A	B	C	D	E
1	Chi-Square Equal Expected Frequencies				
2					
3		fo	fe	ChiSqDst	
4	Cherry	32	=AVERAGE(fo)	=(fo-fe)^2/fe	
5	Strawberry	28			
6	Orange	16			
7	Lime	14			
8	Grape	10			
9					

13. Highlight C4:D4. Place the mouse arrow on the lower right handle of D4. It will look like a thick black plus sign. Click and drag from D4:D8. This copies both the expected frequency and the formula and completes the table . Also, the remaining cells are computed and the output is complete.

Bold cell B11 and B15. The results can now be interpreted.

	A	B	C	D
1	Chi-Square Equal Expected Frequencies			
2				
3		fo	fe	ChiSqDst
4	Cherry	32	20	7.2
5	Strawberry	28	20	3.2
6	Orange	16	20	0.8
7	Lime	14	20	1.8
8	Grape	10	20	5
9				
10	k	5		
11	Chi_Sq	**18**		
12	df	4		
13	Alpha	0.05		
14	Crit_Val	9.487728		
15	Conclusior	**Reject Ho**		
16				

Reject Ho. The computed value for chi-square does not fall in the acceptance range because 18 is more than the critical value of 9.488 (rounded). There is enough evidence to reject the claim that consumers show no preference for the flavors

Save your file as **chi-sqeq 5 cat**

Note: If the expected frequencies are given, key the appropriate values in place of the average of the observed frequencies.

Goodness of fit, example for varied categories

You can use this worksheet to solve other problems with five categories and for problems not having seven categories. You must first re-key the category data in columns A and B. If you have less than 5 categories, you need to clear any cells in rows below the last category. If you have more than 5 categories you must copy the formulas for the extra rows in columns C and D. In both instances you then need to use the Name-Create function to rename columns B, C, and D using the new ranges. When you are asked if you would like to replace the existing definition, click on Yes You may also need to key in a new value for Alpha.

Problem 11-2. The advisor of an ecology club at a large college believes that the group consists of 10% freshmen, 20% sophomores 40% juniors, and 30% seniors. The membership for the club this year consisted of 14 freshmen, 19 sophomores, 51 juniors, and 16 seniors. At $\alpha = 0.10$, test the advisor's conjecture.

1. Open the file **chi-sqeq 5 cat** if not already open.

2. In cells A4:A7, key **Freshmen, Sophomore, Junior**, and **Senior** respectively.

3. In cells B4:B7, key **14**, **19**, **51**, and **16** respectively.

4. Highlight A8:D8. Press the <Delete> key.

Since the expected frequencies are a percentage of the observed frequencies you will enter the formula for finding the expected frequency from a percentage. If the expected frequencies were given, you would type the numbers directly into the column.

5. Highlight C4. Double click your mouse. Place your cursor directly in front of (fo). Backspace to erase the word AVERAGE. Key **.10*SUM** Touch the <ENTER> key.

6. Highlight C5. Double click your mouse. Place your cursor directly in front of (fo). Backspace to erase the word AVERAGE. Key **.20*SUM** Touch the <ENTER> key.

7. Highlight C6. Double click your mouse. Place your cursor directly in front of (fo). Backspace to erase the word AVERAGE. Key **.40*SUM** Touch the <ENTER> key.

8. Highlight C7. Double click your mouse. Place your cursor directly in front of (fo). Backspace to erase the word AVERAGE. Key **.30*SUM** Touch the <ENTER> key

When you finish the cell contents for A4:D7 will look as follows.

	A	B	C	D	E
1	Chi-Square Equal Expected Frequencies				
2					
3		fo	fe	ChiSqDst	
4	Freshmen	14	=0.1*SUM(fo)	=(fo-fe)^2/fe	
5	Sophmore	19	=0.2*SUM(fo)	=(fo-fe)^2/fe	
6	Junior	51	=0.4*SUM(fo)	=(fo-fe)^2/fe	
7	Senior	16	=0.3*SUM(fo)	=(fo-fe)^2/fe	
8					

9. In cell B13 change the value for Alpha to **.10**

As soon as you are finished redefining Column C and entering the new value for Alpha, the output changes to reflect the new data.

Crit_Val | =CHIINV(Alpha, df)

	A	B	C	D	E
1	Chi-Square Equal Expected Frequencies				
2					
3		fo	fe	ChiSqDst	
4	Freshmen	14	10	1.6	
5	Sophmore	19	20	0.05	
6	Junior	51	40	3.025	
7	Senior	16	30	6.533333	
8					
9					
10	k	4			
11	Chi_Sq	**11.20833**			
12	df	3			
13	Alpha	0.1			
14	Crit_Val	6.251394			
15	Conclusior	**Reject Ho**			
16					

Reject Ho. The computed value for chi-square (11.208) is greater than the critical value (6.251). There is enough evidence to reject the advisors claim.

If you wish, save the file as **chi-sqeq 4 cat**. Close your file.

Contingency Tables

You will create a worksheet to determine expected frequencies when observed frequencies are known. You will also create a table to compute chi-square. You will create a worksheet to solve the following problem.

Contingency table for 3 rows and 3 columns

Problem 11-3. A sociologist wishes to see whether the number of years of college a person has completed is related to his or her place of residence. A sample of 88 people is selected and classified as shown.

Location	No college	Four-year degree	Advanced degree	Total
Urban	15	12	8	35
Suburban	8	15	9	32
Rural	6	8	7	21
Total	29	35	24	88

At $\alpha = 0.05$, can the sociologist conclude that a person's location is dependent on the number of years of college?

1. On a new worksheet, enter the data as shown.

	A	B	C	D	E	F
1	Chi-Square Contingency Tables					
2						
3						
4	Location	No college	4-yr deg.	Adv. Deg.	Total	
5	Urban					
6	Suburban					
7	Rural					
8	Total					GrnTot
9						

2. Highlight A3:E8. Under the Home tab, click the Copy icon.

3. Make A10 your active cell. Under the Home tab, click the Paste icon.

4. Make A17 your active cell. Again under the Home tab, click the Paste icon.

5. In cell D3, key **(Observed)**

6. In cell D10, key **(Expected)**

7. In cell D17, key **(Chi-Square Computations)**

8. In cells A24:A27 and B5:D7, enter the data as shown.

	A	B	C	D	E	F	G
1	Chi-Square Contingency Tables						
2							
3				(Observed)			
4	Location	No college	4-yr deg.	Adv. Deg.	Total		
5	Urban	15	12	8			
6	Suburban	8	15	9			
7	Rural	6	8	7			
8	Total					GrrTot	
9							
10				(Expected)			
11	Location	No college	4-yr deg.	Adv. Deg.	Total		
12	Urban						
13	Suburban						
14	Rural						
15	Total						
16							
17				(Chi-Square Computations)			
18	Location	No college	4-yr deg.	Adv. Deg.	Total		
19	Urban						
20	Suburban						
21	Rural						
22	Total						
23							
24	df						
25	Alpha						
26	Crit_Val						
27	Conclusion						
28							

9. In cell B26, key **=CHIINV(Alpha,df)**

10. In cell B27, key **=IF(Chi_Sq>Crit_Val, "Reject Ho", "Do Not Reject Ho")**

11. Highlight A24:B26. Under the Formulas Tab, select the Create from Selection icon. Select Left Column. Click on OK. If you are asked if you would like to replace the existing definition, click on Yes.

11. Make B8 your active cell. Under the Home tab, select the AutoSum icon. Press Enter.

12. Make B8 your active cell again. Place the mouse arrow on the lower right handle of B8. It will look like a thick black plus sign. Click and drag to D8.

14. Make E5 your active cell. Under the Home tab, select the AutoSum icon.

15. Copy E5 into E6:E8.

16. Highlight E8:F8. Use the Name-Create process to name the cell. Select the Right Column.

	A	B	C	D	E	F
1	Chi-Square Contingency Tables					
2						
3				(Observed)		
4	Location	No college	4-yr deg.	Adv. Deg.	Total	
5	Urban	15	12	8	35	
6	Suburban	8	15	9	32	
7	Rural	6	8	7	21	
8	Total	29	35	24	88	GrnTot
9						

The completed Observed table should look as shown above.

You will now compute the expected frequencies. For each cell in the Expected Frequency table (A10:E15), the formula will be: **Expected frequency =(Row total)(Column total)/Grand total**

17. In cell B12, key =**$E5*B$8/GrnTot.**

This formula multiplies the row total (E5) times the column total (B8) and divides by the grand total, which we named GrnTot. The $ sign in the formula keeps the appropriate row or column constant when it is copied into another cell.

18. Make B12 your active cell. Copy the contents to B13:B14 by dragging the right handle.
19. Highlight B12:B14. Drag the right handle of B14 to C14:D14.

20. Make B15 your active cell. Click the AutoSum icon. Press Enter.

21. Copy B15 to C15:D15.

22. Make E12 your active cell. Click the AutoSum icom. Press Enter.

23. Copy E12 to E13:E15.

The completed Expected table should look as shown below.

	A	B	C	D	E
10				(Expected)	
11	Location	No college	4-yr deg.	Adv. Deg.	Total
12	Urban	11.53409	13.92045	9.545455	35
13	Suburban	10.54545	12.72727	8.727273	32
14	Rural	6.920455	8.352273	5.727273	21
15	Total	29	35	24	88
28					

22. In Cell B19, key **=(B5-B12)^2/B12**

This formula subtracts the expected value from the observed value, squares the difference and divides by the expected value. The formula for chi-square is the sum of all the computed results.

23. Copy B19 to B20:B21.

24. Highlight B19:B21. Drag the right handle of B21 to C21:D21.

25. In E22, key **=SUM(B19:D21)**

26. In F22, key **Chi_Sq**

27. Highlight E22:F22. Use the Name-Create process to name the cell. Select the Right column.

The completed Chi-Square Computation table should appear as the table shown below.

	A	B	C	D	E	F
17				(Chi-Square Computations)		
18	Location	No college	4-yr deg.	Adv. Deg.	Total	
19	Urban	1.04148	0.264944	0.250216		
20	Suburban	0.61442	0.405844	0.008523		
21	Rural	0.122425	0.014858	0.282828		
22	Total				3.005539	Chi_Sq
23						

The last thing you will do is enter the values for degrees of freedom and Alpha. The degrees of freedom is df = (number of rows - 1)(number of columns - 1). Or (3-1)(3-1) = 4.

28. In B24, key **4**

29. In B25, key **.05**

Bold cells E22 and B27. This completes your total worksheet and gives the resulting output. The results can now be interpreted.

	A	B	C	D	E	F
1	Chi-Square Contingency Tables					
2						
3				(Observed)		
4	Location	No college	4-yr deg.	Adv. Deg.	Total	
5	Urban	15	12	8	35	
6	Suburban	8	15	9	32	
7	Rural	6	8	7	21	
8	Total	29	35	24	88	GrnTot
9						
10				(Expected)		
11	Location	No college	4-yr deg.	Adv. Deg.	Total	
12	Urban	11.53409	13.92045	9.545455	35	
13	Suburban	10.54545	12.72727	8.727273	32	
14	Rural	6.920455	8.352273	5.727273	21	
15	Total	29	35	24	88	
16						
17				(Chi-Square Computations)		
18	Location	No college	4-yr deg.	Adv. Deg.	Total	
19	Urban	1.04148	0.264944	0.250216		
20	Suburban	0.61442	0.405844	0.008523		
21	Rural	0.122425	0.014858	0.282828		
22	Total				**3.005539**	Chi_Sq
23						
24	df	4				
25	Alpha	0.05				
26	Crit_Val	9.487728				
27	Conclusior	**Do Not Reject Ho**				
28						

The null hypothesis is not rejected at the .05 level of significance. The computed value for chi-square, 3.01 (rounded) is less than the critical value of 9.488 (rounded). There is not enough evidence to support the claim that a person's place of residence is dependent on the number of years of college completed.

Save your file as **chi-sqcn 3r 3c**

Contingency Tables for Varied Rows and Columns

To use this worksheet to solve problems with other than 3 rows and 3 columns, you must modify the worksheet. If you insert or delete rows and columns **between** existing rows and columns, you will not have to re-compute the existing row and column sums, they will adjust automatically. If you insert rows or columns you will need to copy existing formulas to the inserted cells. You will also need to enter the new value for the degrees of freedom.

Problem 11-4. According to a recent survey, 64 % of Americans between the ages of 6 and 17 cannot pass a basic fitness test. A physical education instructor wishes to determine if the percentages of such students in different schools in his school district are the same. He administers a basic fitness test to 120 students in each of four schools. The results are shown here. At $\alpha = 0.05$, test the claim that the proportions who pass the test are equal. (Textbook Exercise 11-3, Problem 23 for demonstration purposes).

	Southside	West End	East Hills	Jefferson
Passed	49	38	46	34
Failed	71	82	74	86
Total	120	120	120	120

1. Open the file **chi-sqcn 3r3c** if it is not already open.

First you will add a column and delete some rows.

2. Make D1 your active cell. Under the Home tab, click the drop down arrow under Insert. Select Insert Sheet Columns.

This inserts a blank column between two existing categories. As you key in the extra data you will notice that the totals in the rows and columns will automatically change to reflect the current data. You will now delete all the rows with the row heading, Suburban.

3. Make A6 your active cell. Under the Home tab, click the drop down arrow under Delete. Select Delete Sheet Rows.

4. Delete row 12. Then delete row 18. (All data rows labeled Suburban)

You deleted a row between two existing rows; therefore, you did not have to re-compute the formulas for totals. You now need to change the row and column headings.

5. In cells A4:A6 key the row headings **Results**, **Passed** and **Failed** respectively.

6. Highlight cells A4:A6. Under the Home tab, click on the Copy icon.

7. Make cell A10 your active cell. Under the Home tab, click on the Paste icon.

8. Make cell A16 your active cell. Under the Home tab, click on the Paste icon.

9. In cells B4:E4, key the column headings **Southside**, **West End**, **East Hills** and **Jefferson**, respectively.

10. Highlight cells B4:E4. Under the Home tab, click on the Copy icon

11. Make cell B10 your active cell. Under the Home tab, click on the Paste icon.

12. Make cell B16 your active cell. Under the Home tab, click on the Paste icon.

At this point your worksheet will look as shown on the following page.

	A	B	C	D	E	F	G
1	Chi-Square Contingency Tables						
2							
3					(Observed)		
4	Results	Southside	West End	East Hills	Jefferson	Total	
5	Passed	15	12		8	35	
6	Failed	6	8		7	21	
7	Total	21	20		15	56	GrnTot
8							
9					(Expected)		
10	Results	Southside	West End	East Hills	Jefferson	Total	
11	Passed	13.125	12.5		9.375	35	
12	Failed	7.875	7.5		5.625	21	
13	Total	21	20		15	56	
14							
15					(Chi-Square Computations)		
16	Results	Southside	West End	East Hills	Jefferson	Total	
17	Passed	0.267857	0.02		0.201667		
18	Failed	0.446429	0.033333		0.336111		
19	Total					**1.305397**	Chi_Sq
32							

Now you will enter the data for problem 4 in cells A5:B6.

12. In cell B5:E5, key **49**, **38**, **46**, and **34** respectively.

13. In cells B6:E6, key **71**, **82**, **74**, **86** respectively.

Notice how most of the cells changed automatically.

You now need to copy the formulas to the blank cells.

14. Copy C7 to D7.

15. Highlight C11:C13. Drag the right handle of C13 to D13.

16. Highlight C17:C18. Drag the right handle of C18 to D18.

The only thing left to do is to put in a new value for the degrees of freedom. Since there are now 2 rows and 4 columns, the degrees of freedom is (2-1)(4-1) = 3.

17. In **B21**, key **3**

Since Alpha is the same (.05) you are finished with the problem.

The output in the worksheet has changed to reflect the changed data.

	A	B	C	D	E	F	G
1	Chi-Square Contingency Tables						
2							
3					(Observed)		
4	Results	Southside	West End	East Hills	Jefferson	Total	
5	Passed	49	38	46	34	167	
6	Failed	71	82	74	86	313	
7	Total	120	120	120	120	480	GrnTot
8							
9					(Expected)		
10	Results	Southside	West End	East Hills	Jefferson	Total	
11	Passed	41.75	41.75	41.75	41.75	167	
12	Failed	78.25	78.25	78.25	78.25	313	
13	Total	120	120	120	120	480	
14							
15					(Chi-Square Computations)		
16	Results	Southside	West End	East Hills	Jefferson	Total	
17	Passed	1.258982	0.336826	0.432635	1.438623		
18	Failed	0.671725	0.179712	0.230831	0.767572		
19	Total					**5.316906**	Chi_Sq
20							
21	df	3					
22	Alpha	0.05					
23	Crit_Val	7.814725					
24	Conclusior	**Do Not Reject Ho**					
25							

Do not reject Ho since Chi square (5.317) rounded is less than the critical value of 7.815 (rounded).

If you wish, save your file as **chi-sqcn 2r 4c**. Close your file.

Practice Exercises taken from textbook.

Hint: Be sure to abbreviate to fit labels in one cell.

11-1. A researcher for an automobile manufacturer wishes to see if the ages of automobiles are equally distributed among three categories: less than 3 years old, 3 to 7 years old, and 8 years or older. A sample of 30 adult automobile owners is selected, and the results are shown. At $\propto = 0.05$ can it be considered that the ages of the automobiles are equally distributed among the three categories? (Textbook Exercise 11-2, Problem 5).

Hint: Use the file chi-sqeq 5 cat. Delete the bottom 2 rows of data. Key in new labels and key in new values for *fo* and *Alpha*.

Category	Less than three years	3 to 7 years	8 years or older
Number	8	10	12

11-2. A researcher wishes to see if the five ways (drinking decaffeinated beverages, taking a nap, going for a walk, eating a sugary snack, other) people use to combat midday drowsiness are equally distributed among office workers. A sample of 60 office workers is selected, and the following data are obtained. At $\propto = 0.10$, can it be concluded that there is no preference? (Textbook Exercise 11-2, Problem 6)

Hint: Use chi-sqeq 5 cat. Key in new labels and key in new values for *fo* and *Alpha*.

Method	Beverage	Nap.	Walk.	Snack	Other.
Number	21	16	10	8	5

11-3. According to a recent census report, 68% of families have two parents present, 23% have only a mother present, 5% have only a father present, and 4% have no parent present. A random sample of families from a large school district revealed these results. . At $\propto$ 0.05, is there sufficient evidence to conclude that the proportions of families by type of parent(s) present differ from those reported by the census?. (Textbook Exercise 11-2, Problem 8).

Hint: Use chi-sqeq 4 cat. Key in new labels, new values for *fo* and *Alpha* and change the percentages for the expected observations as in problem 11-2 in the Excel® workbook.

Two parents	Mother only	Father only	No parent
120	40	30	10

11-4. A police investigator read that the reasons why gun sales to applicants were denied were distributed as follows: criminal history of felonies, 75%; domestic violence conviction, 11%; and drug abuse, fugitive, etc, 14%. A sample of applicants in a large study who were refused sales is obtained and is distributed as follows.. At $\propto = 0.10$, can it be concluded that the distribution is as stated?. (Textbook Chapter 11, Review Exercise # 4).

Hint: Use chi-sqeq 4 cat. Delete the bottom row. Key in new labels, new values for *fo* and *Alpha* and change the percentages for the expected observations as in problem 11-2 in the Excel® workbook.

Reason	Criminal history	Domestic violence	Drug abuse etc
Number	120	42	38

11-5. A study is conducted as to whether there is a relationship between joggers and the consumption of nutritional supplements. A random sample of 210 subjects is selected, and they are classified as shown. At $\propto = 0.05$, test the claim that jogging and the consumption of supplements are not related. (Textbook Exercise 11-3, Problem 8).

Hint: Use chi-sqcn 2r 4c. Delete one of the <u>middle</u> columns. Key in new labels, new observed values, and Alpha.

Jogging status	Daily	Weekly	As needed
Joggers	34	52	23
Nonjoggers	18	65	18

11-6. A researcher surveyed 100 randomly selected lawyers in each of four areas of the country and asked them if they had performed *pro bono* work for 25 or fewer hours in the last year. The results are shown here. At $\propto = 0.10$, is there enough evidence to reject the claim that the proportions of those who accepted *pro bono* work for 25 hours or less are the same in each area? (Exercise 11-3, Problem 27).

Hint: Use chi-sqcn 2r 4c. Key in new labels, new observed values, and *Alpha*.

	North	South	East	West
Yes	43	39	22	28
No	57	61	78	72
Total	100	100	100	100

11-7. A car manufacturer wishes to determine whether the type of car purchased is related to the individual's gender. The data obtained from a sample are shown here. At $\propto = 0.01$, is the gender of the purchase related to the type of car purchased? (Textbook Chapter 11, Review Exercise # 6).
Hint: Use chi-sqcn 2r 4c. Key in new labels, new observed values, and Alpha.

Gender of purchaser	Type of vehicle purchased Sedan	Compact	Station wagon	SUV
Male	33	27	23	17
Female	21	34	41	18

11-8. The risk of injury is higher for males as compared to females (57% versus 43%). A hospital emergency room supervisor wishes to determine if the proportions of injuries to males in his hospital are the same for each of four months. He surveys 100 injuries treated in his ER for each month. The results are shown here. At $\propto = 0.05$, can he reject the claim that the proportions of injuries for males are equal for each of the four months. (Textbook Chapter 11, Review Exercise # 8)
Hint: Use chi-sqcn 2r 4c. Key in new labels, new observed values, and *alpha*.

	May	June	July	August
Male	51	47	58	63
Female	49	53	42	37
Total	100	100	100	100

CHAPTER 12
ANALYSIS OF VARIANCE

In this chapter we will continue to discuss hypothesis testing by introducing the F distribution. The F distribution is used as the test statistic for several situations. It is used to test whether two samples are from populations having equal variances, and it is also applied when we want to compare more than two population means simultaneously. The simultaneous comparison of several population means is called analysis of variance (ANOVA). In both of these situations, the populations must be normal, and the data must be at least interval-scale. After completing this chapter, you will be able to:

- Use Excel® to set up and organize data into an ANOVA table.
- Use Excel®'s Data Analysis ToolPak to complete single factor analysis of variance, or ANOVA, as well as two factor analysis of variance.

You will continue to use the five steps in hypothesis testing:

1. **State the null and alternative hypothesis** using either formulas or words. The Null Hypothesis (H_o) is always the statement of no significant difference. The Alternative Hypothesis (H_1) is always the statement that there is a significant difference. When direction is stated it is a one-directional test (one-tailed). When direction is not stated it is a two-directional test (two-tailed).

2. **State the level of significance** or the probability that the null hypothesis is rejected when, in fact, it is true.

3. **State the statistical test** you will be using: the z test, t test, f test, chi-square test, etc.

4. **Formulate a decision rule**. Using a picture or curve that estimates the distribution you are testing, show the critical value if you are performing a one-directional test or the upper and lower critical values if you are performing a two-directional test.

5. **Do it**. Show the formula you used and at least the major steps involved. State the results of the hypothesis test in terms of the question using complete sentences and examples.

Single Factor Analysis of Variance

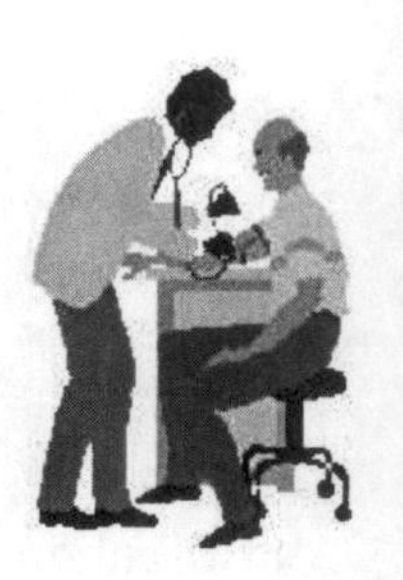

Problem 12-1. A researcher wishes to try three different techniques to lower the blood pressure of individuals diagnosed with high blood pressure. The subjects are randomly assigned to three groups; the first group takes medication, the second group exercises, and the third group follows a special diet. After four weeks, the reduction in each person's blood pressure is recorded. At $\propto = 0.05$, test the claim that there is no difference among the means. The data is as follows.

Medication	Exercise	Diet
10	6	5
12	8	9
9	3	12
15	0	8
13	2	4

	A	B	C
1	Meds	Exercise	Diet
2	10	6	5
3	12	8	9
4	9	3	12
5	15	0	8
6	13	2	4
7			

1. In a new worksheet, key the data as shown.

Remember: The Analysis ToolPak option must be installed and selected before proceeding as noted Chapter 2 instructions.

2. Under the Data tab, click Data Analysis. Select ANOVA: Single Factor. Click OK.

The dialog box for ANOVA: Single Factor is displayed.

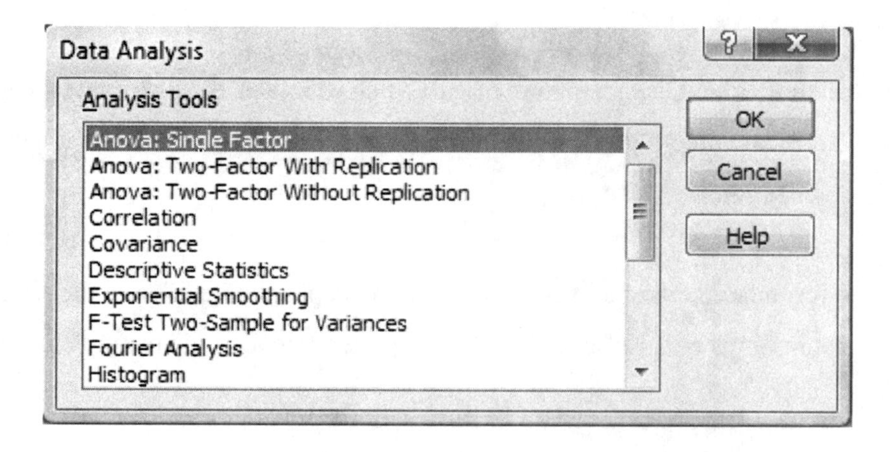

3. Your cursor should be on the Input Range text box. Key **A1:C6**. Press the tab key.

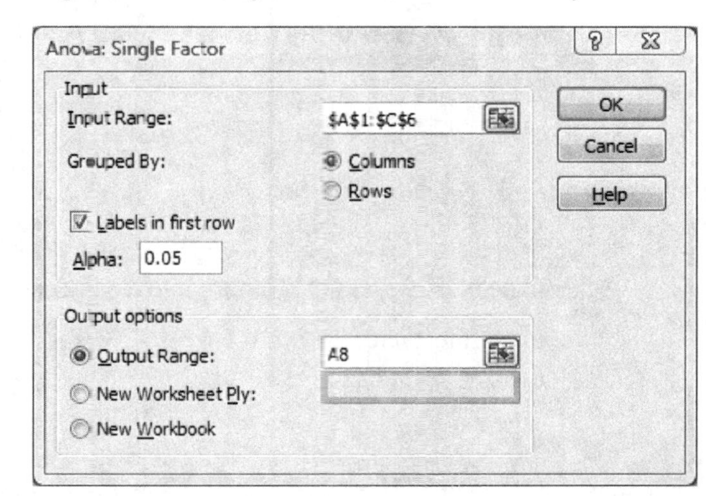

4. Grouped by Columns should be selected. Press the tab key.

5. Select Labels in First Row, check box. Press the tab key.

6. The Alpha text box should have .05

7. Select the radio button for Output Range. In the Output Range text box, key **A8**. Click OK.

The output box is now displayed but it is difficult to read all of the cells.

	A	B	C	D	E	F	G
1	Meds	Exercise	Diet				
2	10	6	5				
3	12	8	9				
4	9	3	12				
5	15	0	8				
6	13	2	4				
7							
8	Anova: Single Factor						
9							
10	SUMMARY						
11	*Groups*	*Count*	*Sum*	*Average*	*Variance*		
12	Meds	5	59	11.8	5.7		
13	Exercise	5	19	3.8	10.2		
14	Diet	5	38	7.6	10.3		
15							
16							
17	ANOVA						
18	*rce of Varia*	*SS*	*df*	*MS*	*F*	*P-value*	*F crit*
19	Between G	160.1333	2	80.06667	9.167939	0.003831	3.88529
20	Within Gro	104.8	12	8.733333			
21							
22	Total	264.9333	14				
23							

8. Make A18 your active cell. Under the Home tab, click the drop down arrow under Format. Select AutoFit Column Width.

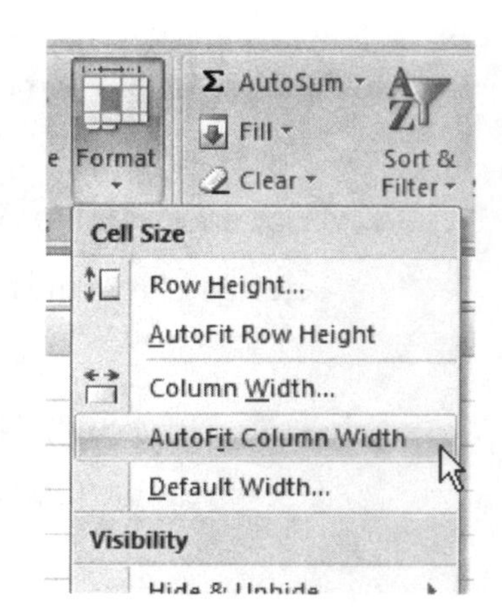

9. Bold cells E19 and G19. Your table is now easier to read and can be interpreted.

The computed F of 9.17 (rounded) is greater than the F critical value of 3.89 (rounded) so we reject the null hypothesis. There is a difference among the means.

G19 fx 3.88529031170037

	A	B	C	D	E	F	G
1	Meds	Exercise	Diet				
2	10	6	5				
3	12	8	9				
4	9	3	12				
5	15	0	8				
6	13	2	4				
7							
8	Anova: Single Factor						
9							
10	SUMMARY						
11	*Groups*	*Count*	*Sum*	*Average*	*Variance*		
12	Meds	5	59	11.8	5.7		
13	Exercise	5	19	3.8	10.2		
14	Diet	5	38	7.6	10.3		
15							
16							
17	ANOVA						
18	*Source of Variation*	*SS*	*df*	*MS*	*F*	*P-value*	*F crit*
19	Between Groups	160.1333	2	80.06667	**9.167939**	0.003831	**3.88529**
20	Within Groups	104.8	12	8.733333			
21							
22	Total	264.9333	14				
23							

If you wish, save your file as **Ch12-prob1**. Close your file.

Two Factor Analysis of Variance

Problem 12-2. A researcher wishes to see whether the type of gasoline used and the type of automobile driven have any effect on gasoline consumption. Two types of gasoline, regular and high-octane, will be used, and two types of automobiles, two-wheel and four-wheel drive, will be used in each group. There will be two automobiles in each group, for a total of eight automobiles used. The data (in miles per gallon) are shown here. Use a two-way analysis of variance with $\alpha = 0.05$.

Type of automobile

Gas	Two-wheel	Four-wheel
Regular	26.7	28.6
	25.2	29.3
High-octane	32.3	26.1
	32.8	24.2

Test the following claims.

a). There is no difference between the means of the gasoline consumption for the two types of gasoline.

b). There is no difference between the means of the gasoline consumption for the two-wheel-drive and the four-wheel-drive automobiles.

c). There is no interaction effect between the type of gasoline used and the type of automobile a person drives on gasoline consumption.

1. In a new worksheet, key the data as shown.

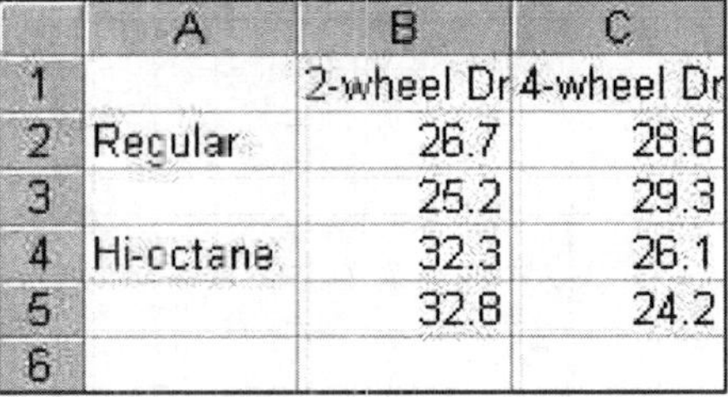

	A	B	C
1		2-wheel Dr	4-wheel Dr
2	Regular	26.7	28.6
3		25.2	29.3
4	Hi-octane	32.3	26.1
5		32.8	24.2
6			

2. Under the Data tab, select Data Analysis. Select ANOVA: Two-Factor With Replication. Click OK.

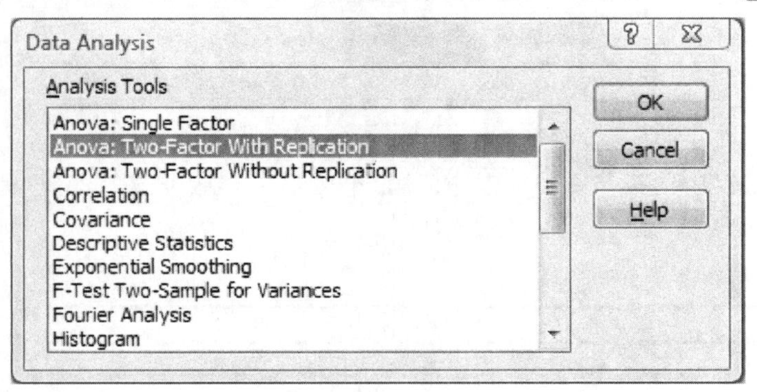

3. Your cursor should be on the Input Range text box. Key **A1:C5**. Press the tab key.

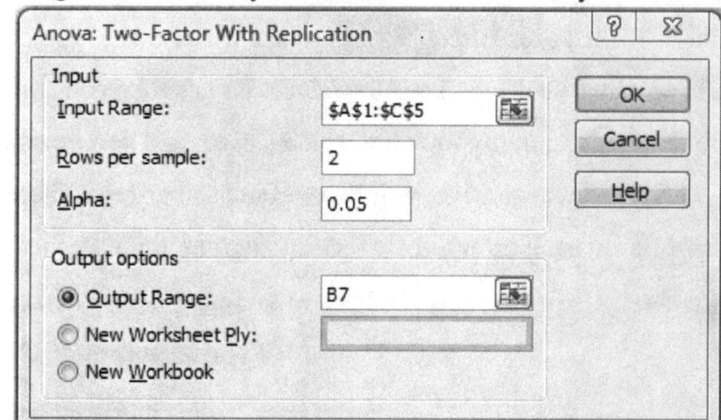

4. Your cursor should be on the Rows per sample text box. Key **2**. Press the tab key.

5. The Alpha text box should have .05

6. Select the radio button for Output Range. In the Output Range text box, key **B7**. Click OK.

	A	B	C	D	E	F	G	H
7		Anova: Two-Factor With Replication						
8								
9		SUMMAR'	2-wheel Dr	4-wheel Dr	Total			
10		*Regular*						
11		Count	2	2	4			
12		Sum	51.9	57.9	109.8			
13		Average	25.95	28.95	27.45			
14		Variance	1.125	0.245	3.456667			
15								
16		*Hi-octane*						
17		Count	2	2	4			
18		Sum	65.1	50.3	115.4			
19		Average	32.55	25.15	28.85			
20		Variance	0.125	1.805	18.89667			
21								
22		*Total*						
23		Count	4	4				
24		Sum	117	108.2				
25		Average	29.25	27.05				
26		Variance	14.93667	5.496667				
27								
28								
29		ANOVA						
30		*Source of Varia*	*SS*	*df*	*MS*	*F*	*P-value*	*F crit*
31		Sample	3.92	1	3.92	**4.751515**	0.094766	**7.70865**
32		Columns	9.68	1	9.68	**11.73333**	0.026648	**7.70865**
33		Interaction	54.08	1	54.08	**65.55152**	0.001265	**7.70865**
34		Within	3.3	4	0.825			
35								
36		Total	70.98	7				
37								

Bold cells F31:F33 and H31:H33. Your table is displayed and can be interpreted.

Since the computed F value for Sample (4.752 rounded) is less than the critical value of 7.71(rounded) we do not reject the null hypothesis for (a). At the .05 level there is no difference between the means of the gasoline consumption for the two types of gasoline.

Since the computed F values for Columns and Interaction (11.733 and 65.552 rounded) are greater than the critical value of 7.71 (rounded), we reject the null hypothesis for b and c. At the .05 level it can be concluded that the type of automobile driven and the combination of type of gasoline and automobile does affect gasoline consumption.

If you wish, save your file as **Ch12-prob2**. Close your file.

Practice Exercises taken from textbook. **Hint:** Abbreviate if you need, to assure all labels fit in one cell.

12-1. The data represent the lengths in feet of three types of bridges in the United States. At $\alpha = 0.01$, test the claim that there is no significant difference in the means of the lengths of the types of bridges. (Textbook Chapter 12, Review Exercise # 1)

Simple truss	Segmented concrete	Continuous plate
745	820	630
716	750	573
700	790	525
650	674	510
647	660	480
625	640	460
608	636	451
598	620	450
550	520	450
545	450	425
534	392	420
528	370	360

12-2. Three random samples of times (in minutes) that commuters are stuck in traffic are shown. At $\alpha = 0.05$, is there a difference in the mean times among the three cities? (Textbook Exercise 12-3, Problem 18)

Dallas	Boston	Detroit
59	54	53
62	52	56
58	55	54
63	58	49
61	53	52

12-3. The number of grams of fat per serving for three different kinds of pizza from several manufactures is listed below. At the 0.01 level of significance, is there sufficient evidence that a difference exists in mean fat content? (Textbook Chapter 12, Review Exercise # 4)

Cheese	Pepperoni	Supreme/Deluxe
18	20	16
11	17	27
19	15	17
20	18	17
16	23	12
21	23	27
16	21	20

12-4. The data consist of the weights in ounces of three different types of digital camera. Use $\propto$ =0.05 to see if the means are equal.? (Textbook Chapter 12, Review Exercise # 6)

2-3 Megapixels	4-5 Megapixels	6-8 Megapixels
6	14	19
8	11	27
7	15	21
11	24	23
4	17	24
8	10	33

12-5. A company wishes to test the effectiveness of its advertising. A product is selected, and two types of ads are written, one is serious and one is humorous. Also the ads are run on both television and radio. Sixteen potential customers are selected and assigned randomly to one of four groups. After seeing or listening to the ad, each customer is asked to rate its effectiveness on a scale of 1 to 20. Various points are assigned for clarity, conciseness, etc. The data are shown here. At $\propto = 0.01$, analyze the data using a two-way ANOVA. (Textbook Exercise 12-4, Problem 10)

Medium

Type of ad	Radio				Television			
Humorous	6	10	11	9	15	18	14	16
Serious	8	13	12	10	19	20	13	17

12-6. A contractor wishes to see whether there is a difference in the time (in days) it takes two subcontractors to build three different types of homes. At $\propto = 0.05$, analyze the data shown here using a two-way ANOVA. (Textbook Exercise 12-4, Problem 12)

Home Type

Subcontractor	I					II					III				
A	25	28	26	30	31	30	32	35	29	31	43	40	42	49	48
B	15	18	22	21	17	21	27	18	15	19	23	25	24	17	13

12-7. A teacher wishes to test the math anxiety level of her students in two classes at the beginning of the semester. The classes are Calculus I and Statistics. Furthermore, she wishes to see whether there is a difference owing to the students' ages. Math anxiety is measured by the score on a 100-point anxiety test. Use $\propto = 0.10$ and a two-way analysis of variance to see whether there is a difference. Five students are randomly assigned to each group. The data are shown here. (Textbook Chapter 12, Review Exercise # 8)

	Class									
Age	Calculus I					Statistics				
Under 20	43	52	61	57	55	19	20	31	36	24
20 or over	56	55	42	48	61	63	78	67	71	75

12-8. A medical researcher wishes to test the effects of two different diets and two different exercise programs on the glucose level in a person's blood. The glucose level is measured in milligrams per deciliter (mg/dl). Three subjects are randomly assigned to each group. Analyze the data shown here using a two-way ANOVA with $\propto = 0.05$. (Textbook Chapter 12, Review Exercise # 9)

Exercise	**Diet**					
program	A			B		
I	62	64	66	58	62	53
II	65	68	72	83	85	91

CHAPTER 13
NONPARAMETRIC STATISTICS

Chapters 8 and 9 dealt with data that were at least interval-scale, such as weights, incomes, and ages. We conducted a number of tests of hypothesis about a population mean and two or more population means. For these tests it was assumed that the population was normal. In Chapter 11, you were introduced to one type of nonparametric test, the chi-square distribution. This chapter continues with use of nonparametric tests, or a distribution-free hypothesis tests. As with the chi-square distribution, the nonparametric distributions in this chapter are also positively skewed. In this chapter you will learn to:

- Create and use a template to perform the sign test
- Create a template for solving problems using the Wilcoxon Rank Sum Test

You will continue to use the five steps in hypothesis testing:

1. **State the null and alternative hypothesis** using either formulas or words. The null hypothesis (H_o) is always the statement of no significant difference. The alternative hypothesis (H_1) is always the statement that there is a significant difference. When direction is stated it is a one-directional test (one-tailed). When direction is not stated it is a two-directional test (two-tailed).

2. **State the level of significance** or the probability that the null hypothesis is rejected when, in fact, it is true.

3. **State the statistical test** you will be using: the z-test, t-test, f-test, chi square test, etc.

4. **Formulate a decision rule**. Using a picture or curve that estimates the distribution you are testing, show the critical value if you are performing a one-directional test or the upper and lower critical values if you are performing a two-directional test.

5. **Do it**. Show the formula you used and at least the major steps involved. State the results of the hypothesis test in terms of the question using complete sentences and examples.

The Sign Test

Example 13-1. Creating a template for solving problems using the Sign Test.

The formula for finding the z-value in the Sign Test is $z = \frac{(X + 0.5) - (n/2)}{\sqrt{n/2}}$ where,

X is the smaller number of positive or negative outcomes in the sample size. It is referred to as *SmNoX.*

n is the sample size. It will be referred to as *n.*

1. Retrieve the file **1sa-mean** created in Chapter 8.

2. In A1, key **Test of Hypotheses**

3. In A2, key **Sign Test**

You need to delete the rows containing the cell names so you can create new names.

4. Highlight A5:A6. Under the Home tab, click the drop down arrow beside the Delete icon. Select Delete Sheet Rows.

5. In cell A6, key **SmNoX.** This will replace the label of StdDev.

6. In B9, key **=((SmNoX+0.5)-(n/2))/(SQRT(n)/2)**

This computes the z value for this test.

7. Highlight A6:B6. Under the Formulas tab, click Create from Selection. Select the check box for Left Column. Click on OK

Some cells will read #DIV?0! or #NUM!. They will be filled in as you find the values for the variables.

z		f_x =((SmNoX+0.5)-(n/2))/(SQRT(n)/2)		
	A	B	C	D
1	Test of Hypotheses			
2	Sign Test			
3				
4	Input Data			
5	n			
6	SmNoX			
7	Alpha			
8	Calculated Value			
9	z	#DIV/0!		
10	Test for Left-Tail			
11	LftCrt_zVal	#NUM!		
12	Conclusion	#DIV/0!		
13	p-value	#DIV/0!		
14	Test for Right-Tail			
15	RtCrt_zVal	#NUM!		
16	Conclusion	#DIV/0!		
17	p-value	#DIV/0!		
18	Test for Two-Tail			
19	AbsCrt_zVal	#NUM!		
20	Conclusion	#DIV/0!		
21	p-value	#DIV/0!		
22				

Save your file as **sign test**

Problem 13-1. Based on past experience, a manufacturer claims that the median lifetime of a rubber washer is at least 8 years. A sample of 50 washers showed that 21 lasted more than 8 years. At $\propto = 0.05$, is there enough evidence to reject the manufacturer's claim?

1. Open the file **sign test** if it is not already open.

2. In cell B5, key **50**, the sample number.

3. In cell B6, key **21**

This is the smaller number, since n-21 = 29.

4. In cell B7 key **.05**, alpha for this problem.

Bold cells B9 and B12.

This completes the problem.

Since the z value of .-99 (rounded) is greater than the left critical value of –1.65 (rounded), do not reject the null hypothesis. There is not enough evidence to reject the claim that the median lifetime of the washers is a least 8 years.

If you wish, save your file as **Ch13-prob1**

	A	B	C
1	Test of Hypotheses		
2	Sign Test		
3			
4	Input Data		
5	n	50	
6	SmNoX	21	
7	Alpha	0.05	
8	Calculated Value		
9	z	**-0.98995**	
10	Test for Left-Tail		
11	LftCrt_zVal	-1.64485	
12	Conclusion	**Do Not Reject Ho**	
13	p-value	0.161099	
14	Test for Right-Tail		
15	RtCrt_zVal	1.644853	
16	Conclusion	Do Not Reject Ho	
17	p-value	0.838901	
18	Test for Two-Tail		
19	AbsCrt_zVal	1.959961	
20	Conclusion	Do Not Reject Ho	
21	p-value	0.322199	
22			

The Wilcoxon Rank Sum Test

Example 13-2. Creating a template for solving problems using the Wilcoxon Rank Sum Test.

To solve problems for the Wilcoxon Rank Sum Test several things must be done. First you will make some changes to the 1sa-mean template you created in chapter 8 to create a new template. Then you will use Excel® to help you rank the raw data you need to use to solve the formula.

The formula for computing the z-value for the Wilcoxan Rank Sum Test is $z = \dfrac{R - \mu_R}{\sigma_R}$

R is the sum of the ranks for the smaller sample size (n_1), it will be referred to as sum.

The formula for finding $\mu_R = \dfrac{n_1(n_1 + n_2 + 1)}{2}$, it will be referred to as mu.

The formula for finding $\sigma_R = \sqrt{\dfrac{n_1 n_2(n_1 + n_2 + 1)}{12}}$, referred to as sigma, where,

n_1 is the smaller of the sample sizes, it will be referred to as n_1

n_2 is the larger of the sample sizes, it will be referred to as n_2.

If both samples are the same size, either size can be used as n_1.

1. Retrieve the file 1samean

2. In A1, key **Test of Hypotheses**

3. In A2, key **Wilcoxon Rank Sum Test**

You need to delete the rows containing the cell names so you can create new names.

4. Highlight A5:A8. Under the Home tab, click the drop down arrow beside the Delete icon. Select Delete Sheet Rows.

5. Keeping the cells highlighted, click the drop down arrow beside the Insert icon. Select Insert Sheet Rows. Click the Align Right icon.

6. Highlight A11. Click the drop down arrow beside the Delete icon. Select Delete Sheet Rows.

7. Keeping the cell highlighted, click the drop down arrow beside the Insert icon. Select Insert Sheet Rows. Click the Align Right icon.

8. Click on cell A8. Using the above method insert a row.

9. In A5:A9, key **n_1**, **n_2**, **mu**, **sigma**, and **sum** respectively.

10. In A12, key **z**

11. In B7, key **=(n_1*(n_1+n_2+1)/2)**

12. In B8, key
    =SQRT((n_1*n_2*(n_1+n_2+1))/12)

13. In cell B12, key =**(sum-mu)/sigma**

These cells will all read #NAME?

14. Highlight A5:B9. Under the Home tab, click Create from Selection. Select the check box for Left Column. Click OK.

15. Highlight A12:B12. Under the Home tab, click Create from Selection. Select the check box for Left Column. Click OK. If you are asked if you want to replace existing definition of 'z', click Yes.

	A	B
1	Test of Hypotheses	
2	Wilcoxon Rank Sum Test	
3		
4	Input Data	
5	n_1	
6	n_2	
7	mu	0
8	sigma	0
9	sum	
10	Alpha	
11	Calculated Value	
12	z	#DIV/0!
13	Test for Left-Tail	
14	LftCrt_zVal	#NUM!
15	Conclusion	#DIV/0!
16	p-value	#DIV/0!
17	Test for Right-Tail	
18	RtCrt_zVal	#NUM!
19	Conclusion	#DIV/0!
20	p-value	#DIV/0!
21	Test for Two-Tail	
22	AbsCrt_zVal	#NUM!
23	Conclusion	#DIV/0!
24	p-value	#DIV/0!
25		

Some cells will read #DIV/0! or #NUM!. They will be filled in as you find the values for the variables.

The cell contents of A5-B12 will look as follows.

	A	B	C	D	E
5	n_1				
6	n_2				
7	mu	=(n_1*(n_1+n_2+1)/2)			
8	sigma	=SQRT((n_1*n_2*(n_1+n_2+1))/12)			
9	sum				
10	alpha				
11	Calculated Value				
12	z	=(sum-mu)/sigma			
13					

Save your file as **wilcox-rank**

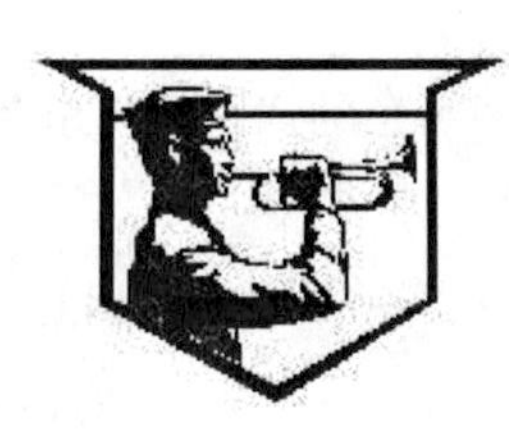

Problem 13-2. Two independent samples of army and marine recruits are selected, and the time in minutes it takes each recruit to complete an obstacle course is recorded as shown in the table. At $\alpha = 0.05$, is there a difference in the times it takes the recruits to complete the course?

Army	15	18	16	17	13	22	24	17	19	21	26	28
Marines	14	9	16	19	10	12	11	8	15	13	25	

Army Mean = 19.67

Marines Mean = 14.24

Using Excel® to rank data

Now you will use Excel® to help rank the data so you can complete the problem.

1. Open the file **wilcox-rank**, if it is not already open.

2. At the bottom of your same worksheet, click on the Sheet 2 tab. This will give you a clean sheet to rank your data.

Sheet1 / Sheet2 / Sheet3

3. In cell A1, key **A** (for Army). In cell A13, key **M** (for Marines).

4. To copy the contents in A1 to A2:A12, make A1 your active cell. Put your mouse pointer on the lower right corner of the cell. It will show a small black box called a handle. The mouse pointer on the handle will show as a thick, black plus. Click your mouse button, hold and drag the mouse pointer down to cell A12.

5. Copy the contents in A13 to A14:A23.

6. In B1:B12, enter the data for Army.

7. In B13:B23, enter the data for Marines.

	A	B
1	A	15
2	A	18
3	A	16
4	A	17
5	A	13
6	A	22
7	A	24
8	A	17
9	A	19
10	A	21
11	A	26
12	A	28
13	M	14
14	M	9
15	M	16
16	M	19
17	M	10
18	M	12
19	M	11
20	M	8
21	M	15
22	M	18
23	M	25
24		

Your worksheet will look as shown on the previous page.

8. Place your cursor anywhere in column B. Under the Data tab, click on the Sort Ascending icon.

This arranges all the data in order so you can rank it.

	A	B	C
1	M	8	1
2	M	9	2
3	M	10	3
4	M	11	4
5	M	12	5
6	A	13	6
7	M	14	7
8	A	15	8.5
9	M	15	8.5
10	A	16	10.5
11	M	16	10.5
12	A	17	12.5
13	A	17	12.5
14	A	18	14.5
15	M	18	14.5
16	A	19	16.5
17	M	19	16.5
18	A	21	18
19	A	22	19
20	A	24	20
21	A	25	21
22	M	25	22
23	A	28	23
24			

9. In column C, place the rank of each value as discussed in the textbook.

10. Place your cursor anywhere in column A. Under the Data tab, click on the Sort Ascending icon.

This arranges the data back into being grouped by Army or Marines. Since Marines is the smaller number, you want the sum of the ranks of the Marines.

	A	B	C
3	A	16	10.5
4	A	17	12.5
5	A	17	12.5
6	A	18	14.5
7	A	19	16.5
8	A	21	18
9	A	22	19
10	A	24	20
11	A	26	22
12	A	28	23
13	M	8	1
14	M	9	2
15	M	10	3
16	M	11	4
17	M	12	5
18	M	14	7
19	M	15	8.5
20	M	16	10.5
21	M	18	14.5
22	M	19	16.5
23	M	25	21
24			93
25			
26			
27			

Sheet1 Sheet2 Sheet3

11. Highlight C13:C23. Under the Home tab, click on the AutoSum icon.

This gives you the sum of the ranks of the data in the smaller sample and gives you the value of R (sum) that you need to compute the z value.

12. At the bottom of your worksheet, click on the Sheet 1 tab.

13. In cell B5, key **11**, the count of items in the smaller sample.

14. In cell B6, key **12**, the count of items in the larger sample.

Cells B7 and B8, mu and sigma, are computed from their formulas.

15. In cell B9, key **93**, the sum of the ranks you just computed in sheet 2.

16. In cell B12, key **.05**, the alpha in this problem for finding the critical value.

Bold cells B12 and B23.

This completes the problem.

This is a two-tailed test, so we reject the null hypothesis, since –2.400(rounded) is less than –1.96 (rounded).

z   fx =(sum-mu)/sigma

	A	B	C
5	n_1	11	
6	n_2	12	
7	mu	132	
8	sigma	16.24808	
9	sum	93	
10	Alpha	0.05	
11	Calculated Value		
12	z	**-2.40028**	
13	Test for Left-Tail		
14	LftCrt_zVal	-1.64485	
15	Conclusion	Reject Ho	
16	p-value	0.008191	
17	Test for Right-Tail		
18	RtCrt_zVal	1.644853	
19	Conclusion	Do Not Reject Ho	
20	p-value	0.991809	
21	Test for Two-Tail		
22	AbsCrt_zVal	1.959963	
23	Conclusion	**Reject Ho**	
24	p-value	0.016382	
25			

There is enough evidence to support the claim that there is a difference in the times it takes the recruits to complete the course.

If you wish, save your file as **Ch13-prob2**

Practice Exercises taken from textbook.

13-1. One hundred people were placed on a special exercise program. After one month, 58 had lost weight, 12 gained weight, and 30 weighed the same as before. Test the hypothesis that the exercise program is effective at $\alpha = 0.10$. **Note**: It will be effective if fewer than 50% of the people did not lose weight. (Textbook Exercise 13-3, Problem 10).

13-2. For a specific year, the median price of natural gas was $10.86 per 1000 cubic feet. A researcher wishes to see if there is enough evidence to reject the claim. Out of 42 households, 18 paid less than $10.86 per 1000 cubic feet for natural gas. Test the claim at $\alpha = 0.05$.
(Textbook Exercise 13-3, Problem 9)

13-3. A researcher read that the median age for viewers of the Carson Daly show is 39. To test the claim, 75 viewers were surveyed, and 27 were under the age of 39. At $\alpha = 0.02$ test the claim.
(Textbook Exercise 13-3, Problem 12)

13-4. A tire manufacturer claims that the median lifetime of a certain brand of truck tires is 40,000 miles. A sample of 30 tires shows that 12 lasted longer than 40,000 miles. Is there enough evidence to reject the claim at $\alpha = 0.05$? (Textbook Chapter 13, Review Exercise # 2)

13-5. A random sample of men and women in prison was asked to give the length of sentence each received for a certain type of crime. A $\alpha = 0.05$, test the claim that there is no difference received by each gender. The data (in months) are shown here. (Textbook Exercise 13-4, Problem 4)

Males	8	12	6	14	22	27	32	24	26	19	15	13		
Females	7	5	2	3	21	26	30	9	4	17	23	12	11	16

13-6. To test the claim that there is no difference in the lifetimes of two brands of handheld video games, a researcher selects a sample of 11 video games of each brand. The lifetimes (in months) of each brand are shown here. At $\alpha = 0.01$, can the researcher conclude that there is a difference in the distribution of lifetimes for the two brands? (Textbook Exercise 13-4, Problem 6)

Brand A	42	34	39	42	22	47	51	34	41	39	28
Brand B	29	39	38	43	45	49	53	38	44	43	32

13-7. Supervisors were asked to rate the productivity of employees on their jobs. A researcher wishes to see whether married men receive higher ratings than single men. A rating scale of 1 to 50 yielded the data shown here. At $\alpha = 0.01$, is there evidence to support this claim? (Textbook Exercise 13-4, Problem 10)

Single men	48	46	42	50	38	36	40	31	28	24	49	34
Married men	44	35	41	37	42	43	29	31	37	32	36	

13-8. Samples of students majoring in business and engineering are selected, and the amount (in dollars) each spent on a required textbook for the fall semester is recorded. The data are shown here. For the Wilcoxon Rank Sum Test at $\alpha = 0.10$, is there a difference in the amount spent by each group? (Textbook Chapter 13, Review Exercise # 5)

Business		Engineering	
48	36	98	73
52	62	72	78
74	50	63	93
63	46	78	88
51	53	55	86
49	58	58	85
		64	

CHAPTER 14
SAMPLING AND SIMULATION

As you have learned previously, sampling is fundamental to the concept of statistics. The advent of computer technology has tremendously improved the ability to quickly generate unbiased subgroups (samples) of a larger population. The Analysis ToolPak in Excel® includes a Random Number Generation analysis tool that allows you to generate random samples from the defined population (input range), as well as stratified random samples (referred to as periodic). After completing this chapter, you will be able to:

- Generate random samples using the Analysis ToolPak
- Compare the mean of random samples with the population mean

Random Number Generation Analysis Tool

You will need to open the Databank file located on the CD that was enclosed with your textbook or located on the publisher's website. Refer to the instructions for accessing the data sets at the end of Chapter 1.

You will use only the Age and Weight columns at this time.

1. At the top of your worksheet, click column heading C and drag to column E.
2. Under the Home tab click the Delete icon.
3. At the top of your worksheet, click column heading E and drag to column I.
4. Under the Home tab click the Delete icon.
5. Save your data on your storage device as **Chapter14 data**
6. In cells D1:I2 enter the cell contents as shown.

	A	B	C	D	E	F	G	H	I
1	ID-NUMBER	AGE	WEIGHT	Mean	Mean	Mean sample Age		Mean Sample Weight	
2	1	27	120	popu. Age	popu. Weight	1	2	1	2
3	2	18	145						
4	3	32	118						
5	4	24	162						
6	5	19	106						

7. In cell D4, key **=AVERAGE(B2:B51)**

8. In cell E4, key **=AVERAGE(C2:C51)**

This gives you the mean of the population Age and Weight.

You will now take a couple of samples from each population and compare the mean with the population mean.

9. Under the Data tab, click Data Analysis. Select Sampling. Click OK.

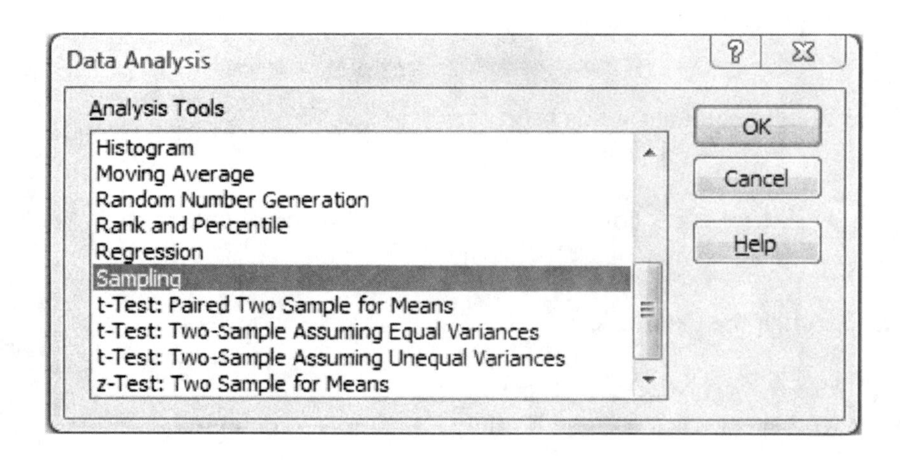

10. In the Input Range textbox, key **B1:B51**

11. Select the checkbox for Labels.

12. The radio button for Random should be selected under Sampling Method.

13. In the Number of Samples textbox, key **10**

16. Select the radio button for Output range.

17. In the Output Range textbox, key **F3**

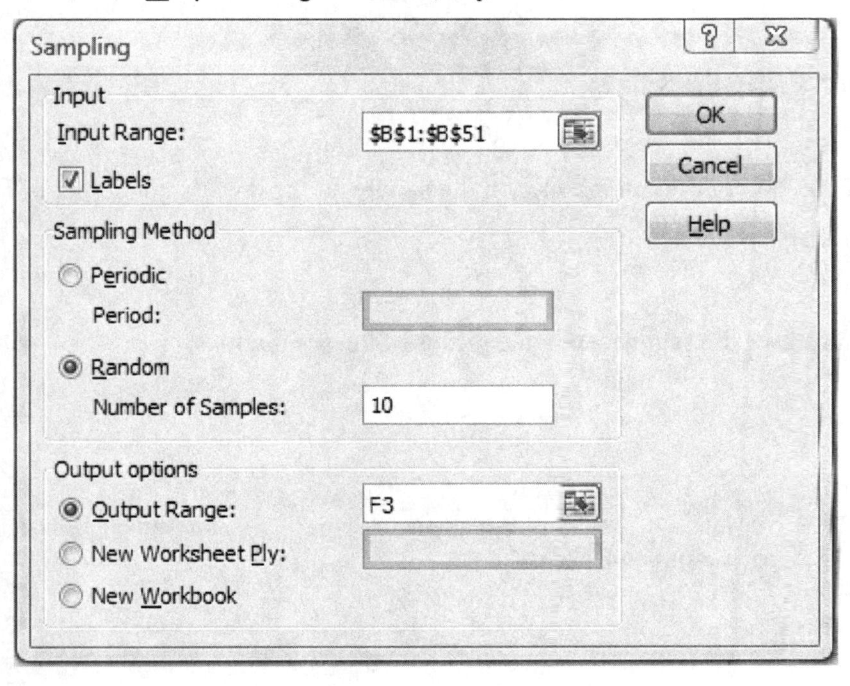

18. Select OK.

19. Repeat selecting another sample of size 10 using, Data Analysis, and Sampling.

All the data should be the same except for the following:

20. In the Output Range textbox, key **G3**. Select OK.

You should now have two samples of ten from the population GPA. You will repeat the steps to take two samples from the IQ population.

21. Select Data Analysis and Sampling.

22. In the Input Range key **C1:C51**

23. In the Output Range, key **H3**

24. Select **Data Analysis** and **Sampling.**

25. In the **Output Range**, key **I3**

26. In F13, key **=AVERAGE(F3:F12)**

27. Click on cell F13. Put your mouse pointer on the handle on the lower right corner of F13. Click, hold and drag to G13:I13.

You can now compare the means of the samples with the means of the populations.

Your worksheet will vary.

	A	B	C	D	E	F	G	H	I
1	ID-NUMBER	AGE	WEIGHT	Mean	Mean	Mean sample Age		Mean Sample Weight	
2	1	27	120	popu. Age	popu. Weight	1	2	1	2
3	2	18	145	39.02	149.46	24	25	162	119
4	3	32	118			44	31	199	148
5	4	24	162			48	59	191	215
6	5	19	106			25	19	112	170
7	6	56	143			28	53	170	101
8	7	65	160			42	55	131	123
9	8	36	215			27	63	121	118
10	9	43	127			35	57	187	143
11	10	47	132			31	48	143	120
12	11	48	196			28	42	149	121
13	12	25	109			33.2	45.2	156.5	137.8
14	13	63	170						

If you wish, save your worksheet as **Ch14-exercise**.

Notes

Notes